Brinjal
(*Solanum melongena* Linn.)

Dr. Kanaya Lal Bhat Professor, Division of Vegetable Science and Floriculture, Sher-e-Kashmir University of Agricultural Sciences and Technology (J & K). Dr. Bhat received his Ph.D. in Vegetable Science from HPKV, Palampur (HP) in 1994. He has taught vegetable science courses to M.Sc and Ph.D students at SKUAST–Kashmir and SKUAST–Jammu for about 20 years and has guided a number of M.Sc and a Ph.D. students in the discipline of vegetable production. He is author of numerous research papers and has written extensively in scientific press and for the general public. He has participated in several national seminars and workshops on vegetable. He has to his credit two books *viz., Minor Vegetables Un-tapped Potential, Physiological Disorders of Vegetable Crops* which has received wide acclaim from the students of agriculture, horticulture and vegetable growers.

Brinjal
(*Solanum melongena* Linn.)

By
K.L. Bhat

2011
DAYA PUBLISHING HOUSE®
Delhi - 110 035

© 2011, KANAYA LAL BHAT (b. 1947–   )
ISBN 9788170359784

Published by	:	**Daya Publishing House** **A Division of** **Astral International Pvt. Ltd.** **– ISO 9001:2008 Certified Company –** 4760-61/23, Ansari Road, Darya Ganj, New Delhi - 110 002 Phone: 23245578, 23244987 Fax: (011) 23260116 e-mail : dayabooks@vsnl.com website : www.dayabooks.com
Laser Typesetting	:	**Classic Computer Services** Delhi - 110 035
Printed at	:	**Chawla Offset Printers** Delhi - 110 052

PRINTED IN INDIA

Preface

India is the second largest producer of vegetables in the world, next only to China with an estimated production of about 105 million tonnes from an area of 6.0 million hectares at an average yield of 16 tonnes per hectare. India shares about 15 per cent of the world output of vegetables from about 2.8 per cent of the cropped area in the country. However, per capita consumption in India makes it possible to grow wide variety of vegetable crops all the year round in one part of the country or another. India can claim to grow the largest number of vegetable crops compared to any other country of the world and as many as 61 annuals and 4 perennials vegetable crops are commercially cultivated. Indians are predominantly vegetarians and depend on vegetables for bulk of their nutrients and minerals. They form the most important component of a balanced diet. The role and the usefulness of the anti-oxidants present in vegetables for human health has been demonstrated recently, therefore, adding value and luster to these crops.

Among the Solanaceous vegetables, brinjal, commonly known as egg plant *Solanum melongena* Linn. is the most common, popular, affordable and principal vegetable crop grown extensively in almost all geographical parts of the country except at higher altitudes. Brinjal being native to India, is cultivated in kitchen gardens and in small family farms for whom it is a source of cash income for small farmers.It occupies the third position amongst vegetable crops. India remains the worlds second producer of brinjal, accounting for nearly 26 per cent of the global production. India's brinjal economy is estimated to be close to $2 billion (9600 crore) with 1.4 million farmers cultivating it annually in nearly 550.000 hectares (1.4 million acres).

Its production will contribute not only to food and nutrition security but also to poverty alleviation and income generation. Its cultivation stimulates development because it is easy to grow, labour incentive, earns higher returns, does not involve extra skills and is available round the year. Brinjal is low in calories, high in nutrition, very high in water content and has therapeutic value. It is very good source of fibre, calcium, phosphorus, folate and vitamin B & C. It is sweet, sharp, hot, cures fevers, excessive humour of phlegm and flatulation. It is appetizer, increases virility and is light. Brinjals of small size is delicate and it cures excessive phlegm and bile in the body and is known to protect arteries from cholesterol damage. Brinjals, if taken in its meshed form or as soup with asafoetida and garlic, it cures flatulation, insomnia and the enlarged spleen due to malaria. Brinjal may be marinated, stuffed, grilled, fried, backed or stir-fried. It is an essential ingredient in several savoury dishes, most commonly *baingan ka bhartha.*

There are hundred's of cultivars of different colours and shapes available in different parts of the country and their number is growing unabatedly. As various Universities, Research Institutions and private sector companies are involved in developing high yielding and disease and pest resistant varieties. The varieties popular in the country are Arka Navneet, Pusa Ankur, Hy 6, Pusa Hybrid-5, ARBH-1, ABH-1, PPL, PPC, Ritu Raj etc. In addition to these there are region specific varieties which are popular among the people of that particular region, because of their colour preference, size or taste etc. Because being cheap as compared to other vegetables, is also known as poorman's vegetable and is available throughtout the year in the country.

Almost 40 per cent of the brinjal produced in India is destroyed by Shoot and Fruit borer. Attacks of *Leucinode orbonalis* is so wide spread that production has barely increased by 9 per cent in the last 10 years despite of 15 per cent larger cultivated area.

This book contains the varied and valuable information on origin, distribution, taxonomy, botany, climate, soil, crop production techniques, nutrition, improved varieties, seed production, breeding and breeders seed production and plant protection measures.The information provided in this book will help both students of Agriculture and the vegetable growers in the production of quality brinjal. This book is therefore, a synthesis of available literature, personal knowledge and the experience shared from the vegetable growers over the years.

My sincere thanks are due to my colleagues for help rendered in various ways in the completion of this book. The author will welcome suggestions and criticisms on the contents of the book for further improvement.

Kanaya Lal Bhat

Contents

and Anther Dehiscence, Pollination, Chromosome Number, Hybridization, Hybrid Vigour, Exploitation of Heterosis, Genetics, Gene Studies in Brinjal, Quantitative, Qualitative Genes, Breeding Objectives.

Punjab Neelam, Punjab Sadabahar, Punjab–8, Pusa Ankur, Pusa
Anmol, Pusa Anupam, Pusa Bindu, Pusa Hybrid–5, Pusa Hybrid
6, Pusa Hybrid-9, Pusa Kranti, Pusa Purple Cluster, Pusa Purple
Long, Pusa Purple Round, Pusa Upkar (DBR-8), Pusa Uttam,
Pusa Bhairav (11a × PPL)-2-4-8-2), Ram Nagar Giant, Rajindra
Baingan, Ravaiyya (MHB-39), RHRBH-1, RHRBH-2,
RHRBH-3, Round-14 (Indo-American Hybrid Seed Co.),
Shyamal (ARBH-201), Surya (SM 6-7), Swetha (SM 6-6), Swarna
Shree (CHES-157), Swarna Mani, Swarna Pratibha, Swarna
Shyamli, S-1, Swarna Shakti, Swarna Ajay, Swarna Shobha,
Surati Gota, Suphal, Type-3, Utkal Jyoti (BB-13), Utkal Keshari
(BB-26), Utkal Madhuri (BB-44), Utakal Tarini (BB-77), Vijay
Hybrid, VRBHR-1, Brinjal Varieties Grouped on the Basis of
Shape and Colour of Fruits, Brinjal Varieties Grouped on the
Basis of their some Morphological Distinctions, Brinjal Hybrid
F1, A. Long Type, B. Round Type, C. Oval to Oblong, D. Small
Sized Fruits.

Chapter 1
Introduction

Brinjal (*Solanum melongena* L.) is one of the most important and widely grown vegetable crop in Asia. Among the solanaceous vegetables, brinjal is extensively grown in India and is very popular among people of all social strata, grown both in home and market gardens throughout the year except at higher altitudes. Brinjal has been cultivated in India for last 4000 years. It is commonly known as Egg plant, is a very common and affordable vegetable. The area under cultivation is estimated to be around 5 lakh hectares and the total production stands at about 82 lakh metric tones. It is mainly grown in small plots as a cash crop by farmers and the average yield in India is around 200–350 quintals per hectare. It is popular in other countries like Japan, Indonesia, Philippines, China and Bulgaria and to some extent in other tropical countries like Africa and America. The major producers of brinjal are China, India,

Japan and Turkey. In India it holds a position nearly comparable to tomato or potato. The ease with which it can be cultivated and its adaptability to a wide range of growing conditions make it popular among the vegetable growers. The nutritional and medicinal properties, make it imperative to grow it all year round. It is used in a variety of culinary preparations. It is highly productive and finds place as the poor man's crop and is liked by poor and the rich. The crop is extremely variable in India. The export of this crop is negligible, and mostly it is consumed locally. It is a delicate perennial often cultivated as an annual. It grows 40–150 cm tall, with large coarsely lobed leaves that are 10–20 cm long and 5–10 cm broad, whereas, semi wild types can grow much larger to 225 cm, with large leaves over 30 cm long and 15 cm broad. The stem is often spiny. The flowers are white to purple, with a five lobed corolla and yellow stamens. The fruit is fleshy, less than 3 cm in diameter on wild plants but much larger in cultivated forms. The fruit is botanically classified as a berry and contains numerous small, soft seeds, which are edible but are bitter because they contain (an insignificant amount of) *nicotinoid* alkaloids, un-surprising as it is a close relative of tobacco. Two crops are typically grown per year in South Asia and because fruits can be harvested every week farmers are provided with an assured income and resource–poor consumers have access to a much needed nutritious vegetable in the summer months when other vegetables are in short supply.

Origin and Distribution

Brinjal is considered to have originated in Indo–Myanmar region (Vavlio, 1928). There are several theories on its origin, which variably consider India, where it was known since ancient times (Bhaduri, 1932) and Baily (1949), Thompson and Kelly (1957), Tsao and Lo (2006), Doijode (2001), whereas De Candole (1886) conceived that egg plant was known to India since ancient times and plant spread towards Africa before the middle ages. Purewal (1957), reported that it is still found growing in India. Martin and Rhodes (1979) reported that a wild form with many small fruits, sometimes called as var. *insanum*, is found on the Bengal plains of India. According to Zeven and Zhukovsky (1975) it originated in India but has a secondary center of variation in China, where it has been known for last 1500 years. Roxburgh (1932) considered *S. melongena* var. *incanum* as a native of Amboyna. Rauwolf had seen this plant in 16th century in a garden in Alleppo. It was called Melanzana and Dedengium. Coulter has also mentioned that according to an herbal of John Gerad, the plant was known in England in 1896. According to ancient *Amer Kosha*, a Sanskrit dictionary of 1100 AD, this plant has been described as Barthaker, Hilgall, Bhand and Duspradarshini. Many authors like Filov, Coulter, and Gazenbus reported that India was the center of origin of *S. melongena*, while a majority of species constituting this genus had originated in South and Central America. India is second largest producer of brinjal in world (Table 1), with an annual production of about 8.5 million tones from an area of 0.50 million hectares. Its cultivation is maximum in Orissa, West Bengal, Bihar and is also distributed in almost all over the country. Various forms, colours and shapes of brinjal are found throughout South East Asia, suggesting that this area is an important center of variation and possibly of origin. Burkill (1935) reported that wild yellow fruited types after the name *Solanum ferox* are found in Malaysia. Sampson (1936) suggested the African origin of this crop, but there is no clear cut evidence that *Solanum melongena* belonged to the area, though there are spiny African brinjal plants. Wild brinjal types are also available in Southern Costal areas of Africa. Brinjal was introduced into Europe, possibly in the thirteenth century and later to Africa via Iran. It has now become widely distributed throughout the tropics.

Brinjal is native to India and was first domesticated over 4000 years ago. It has been cultivated in Southern and Eastern Asia since prehistory, but appears to have become known to the Western world not earlier than ca. 1500 CE. The first known written record of the egg plant is found in Qi min yao shu, an ancient Chinese Agricultural Treatise completed in 544 CE. The migration of brinjal continued in the 9th–12th centuries to the Middle East and Westward to Egypt. The moors introduced the egg plant to *Spaniards* and fruit became popular all over Europe. The Spaniards thought the egg plant was an aphrodisiac and referred to as Berengenas or "the apple of love". This of course added greatly to the popularity of the unusual fruit. The love affair with the brinjal took a downward turn in Northern Europe where Albert of Cologne referred to the fruit as "*Mala Insana*" or Mad Apple (a take-off of the Italian name: "melanzana"). In early years, egg plant also was termed mala insans and the Italian melanzana both of which translate to mad apple. It was a common belief that ingestion of egg plant could cause madness.

It seems the commoners got over that and by the 1600's several varieties migrated from Naples to Germany. While Spainards were traveling the globe they took the brinjal to

South America around 1650. It was Thomas Jefferson (well known for his promotion of Horticulture) who introduced them to the United States in 1806 after receiving an egg plant from a friend in France, from there the cultivation of brinjal spread rapidly all over the United States.

The numerous Arabic and North African names, indicate that it was introduced throughout the Mediterranean area by the Arabs in the early middle ages. The scientific name *Solanum melongena* is derived from a 16th century Arabic term for one type of egg plant. The fruit was introduced to China around 500 B.C. The Chinese hybridized their own varieties of different shaped and coloured egg plants. In China as part of her "bride price" a woman used to be required to produce at least 12 brinjal recipes prior to her wedding day.

The name egg plant is believed to derive from Gerard's description of early forms with small, white fruit resembling eggs, in the United States, Australia, New Zealand and Canada, because the fruits of some 18th century European cultivars were yellow or white and resembled goose or hen's eggs. The name aubergine in British English developed from the French aubergine (as derived from catalan alberginia, from Arabic al-badin jan, from Persian badin–gan, from Sanskrit Vatin-ganah. In India and South Africa English, the fruit is known as brinjal. Aubergin and brinjal with their distinctive br-jn or brn-jil aspects, derive from Arabic and Sanskrit. In caribbean Trinidad, it also goes by the Latin derivative "melin loongen."

The ancestral form was very likely a spiny plant with small, bitter fruit, but selection for improved palatability and for relative spinelessness resulted in gradual emergency of an acceptable type.

Therefore, was so named because early cultivars had egg-shaped fruits. These small egg shaped fruits first grown in Europe, probably were used for commercial purposes about 1500 years ago. The oldest records regarding egg plants are found in a Chinese book written in the 5th century. It is reported to have been introduced into Germany in 1550 and into United States of America in 1806 by Spaniards. Although has never been a major crop in America, but is lately grown to some extent in most areas.

There is a wide diversity in plant habit, vegetative characters, floral morphology and more importantly in fruit size, shape and colour. Hence, local or regional preferences on the basis of fruit characters are very common which account for a large number of varieties in cultivation, including local and imported cultivars. Secondary diversity has been reported in South East Asia and China, while in Africa other species *S. macrocarpon* L. and *S. incanum* L. are important.

Area and Production

Production of brinjal is highly concentrated with 85 per cent of output coming from three countries. China is the top producer (56 per cent of the world output) and India is second (26 per cent); Egypt, Turkey and Indonesia round out the top producing nations. More than 4 million acres (2,043,788 hectares) are devoted to the cultivation of Egg plant in the world. Of all the worlds production, Asia grows 78 per cent and Turky grows 19 per cent, which is the highest production in European Union.

India is one of the major producer of brinjal in the world. It is grown in almost all over the country except cold regions like Leh–Ladakh, because of short growing season. The major brinjal producing states in India are given in Table 4. Apart from South Asia, it is also being grown in China, Turkey, Japan, Egypt, Italy, Indonesia, Iraq, Syria, U.K and Philippines. World production of brinjal is estimated to be about 14.6 million tones (1999). The leading brinjal producing countries (Table 1) in order of importance are China followed by India. In India, brinjal is cultivated in 5 lakh hectares with a total production of about 82 lakh metric tones.

Table 1: Top Ten Brinjal Producing Countries 11ᵗʰ June 2008

Country	Production (Tones)	Foot Note
China	18033,000	F
India	8450,200	
Egypt	1000,000	F
Turkey	791,190	
Indonesia	390,000	F
Iraq	380,000	F
Japan	375,000	F
Italy	271,358	
Sudan	230,000	F
United Kingdom	198,000	F
Total	32,072972–A	

F: FAO estimate.

No symbol: Official figure.

P: Official figure

A: Aggregate (may include, official, semi official, or estimate).

Table 2: Area, Production and Productivity in Different Countries

Country	Area in Ha		Production in Mt		Productivity in kg/ha	
	2003	2004	2003	2004	2003	2004
China	851,627	901,565	16,029,929	16,529,300	18822	18334
Egypt	36,000	36,000	710,000	710,000	19722	19722
India	500,000	510,000	7,830,000	8,200,000	15660	16078
Indonesia	44,414	34,550	301,030	255,359	6777	07391
Italy	12,881	12,355	368,991	362,808	28646	29365
Japan	12,000	11,500	395,000	400,000	32916	34782
Philippines	21,000	21,084	176,999	182,736	8428	8667
Sudan	12,000	12,000	230,000	230,000	19166	19166
Thailand	11,500	11,500	67,000	67,000	5826	5826
Turkey	37,000	37,000	935,000	935,000	25270	25270
World	1,649,167	1,700,655	28,963,901	29,840,793	17562	17546

Table 2a: Worldwide Production of Brinjal

Continent/Country	Production (1000MT)
World	8682
Africa	532
Algeria	25
Egypt	350
Sudan	70
North/Central America	86
Mexico	30
USA	37
Indonesia	175
Iraq	160
Japan	520
Philippines	112
France	27
Greece	71
Italy	261
Spain	130
India	–

Source: Food and Agriculture Organization, Production Year Book, Rome, Italy, 1992

Table 3: Area, Production and Productivity of Brinjal

Year	Area (000'ha)	Production (000'mt)	Productivity (mt/ha)
1987-88	202.6	2552.3	12.6
1991-93	NA	NA	NA
1993-94	300.7	4612.2	15.3
1994-95	420.1	6232.2	14.8
1995-96	434.2	6443.1	14.8
1996-97	464.0	6585.6	14.2
1997-98	486.8	7735.4	15.9
1998-99	496.2	7881.5	15.9

Ref: Horticultural Data Base (2000) NHB, Gurgaon.

Table 4: Statewise Area, Production and Productivity In India (1998-99)

State	Area (1000 ha)	Production (1000 MT)	Productivity (MT)
Andhra Pradesh	79.34	1606.63	20.25
Maharashtra	65.26	1249.00	19.14
Karnataka	49.20	780.80	15.87
Orissa	33.00	405.90	12.30
Madhya Pradesh	30.14	317.37	10.53
West Bengal	18.16	203.39	11.20
Uttar Pradesh	9.27	95.94	10.35
Tamil Nadu	8.84	174.67	19.76
Assam	3.42	28.10	8.25
Rajasthan	3.20	20.47	6.36
Bihar	3.14	21.98	7.00
Gujarat	3.57	37.48	10.50
Punjab	2.80	23.66	8.45
Haryana	2.45	23.59	9.63
Manipur	2.12	12.72	6.00
Kerala	2.10	22.68	10.80
Mizoram	1.33	6.631	4.75
Tripura	1.26	8.341	6.60
Arunachal Pradesh	1.19	5.23	4.40
Pondicherry	0.45	4.86	10.80
Total	320.24	5049.00	15.76

Andhra Pradesh stands first with 24.8 per cent of total area and 31.8 per cent of total production to its credit. The other states namely, Maharashtra, Karnataka, Orissa, Madhya Pradesh and West Bengal comprise about 61.5 per cent of the total area under brinjal.

**Table 5: Area, Production and Productivity of
Major Vegetables in India (2002-2003)**

Vegetables	Area (1000 ha)	Production (1000 MT)	Productivity (MT)
Potato	1337	23161	17.3
Onion	424	4209	9.9
Tomato	478	7616	15.9
Brinjal	507	8001	15.8
Cauliflower	254	4444	17.5
Cabbage	234	5392	23
Green peas	305	2061	6.8
Okra	329	3244	9.9

Areas of Cultivation

Tropical Africa–North Africa, Nigeria, Ghana, Ivory Cost, Kenya and many other countries. Tropical Asia–India, Malaysia, Thailand, Indonesia, Myanmar and Philippines.

Tropical America–Puerto Rico, Haiti, the West Indies and throughout most tropical and subtropical areas.

Current Status

India is second largest producer of vegetables, with an annual production of 87.53 million tonnes from 5.86 million hectares having a share of 14.4 per cent to the world production. Adoption of high yielding cultivars and F1 hybrids and suitable production techniques have largely contributed for higher production and productivity (Table 4). More than 40 kinds of vegetables belonging to different groups, namely cucurbits, cole crops, solanaceous, root and leafy vegetables are grown in different climatic regions of the country. Except a few, brinjal (Egg plant), colocasia, cucumber, ridge gourd, spongy gourd, pointed gourd etc.

most of the other vegetables have been introduced from abroad.

India is the world's second producer of brinjal, accounting for nearly 26 per cent of the global production just behind 30 per cent of China. It occupies the third position amongst vegetable crops. The production in the year 2002-2003 was 80,01,000 MT from 507,000 hactares with an productivity of 15.8 MT. Almost 40 per cent of the brinjal produced in India is destroyed by shoot and fruit borer. Attacks by *Leucinoda orbonalis* is so wide spread that production has barely increased by 9 per cent in the last 10 years despite a 15 per cent larger cultivated area. Yet India's brinjal economy is estimated to be close to $2 billion (Rs. 9600 crores) with 1.4 million small and resource poor farmers cultivating it annually in nearly 550,000 hectares (1.4 million acres). A major factor for stagnant production is the repeated attacks of Fruit Shoot Borer (FSB) which posses a serious problem.

It is due to improved production technology, protection measures and the genetic improvement in yield, quality, diseases and insect-pest resistance that contributed in over all performance of the crop. The yield of long, round, oblong and small round varieties have reached at 50, 65, 60 and 40 t/ha respectively, whereas F1 hybrids raised it to 62.5,78,75 and 50 tonnes per hectare in these respective groups. The varieties of brinjal popular in the country are Arka Navneet, Pusa Ankur, Hybrid 6, Pusa Hybrid-5, ARBH-1, ABH-1, Pusa Purple Long, Pusa Purple Cluster, Ritu Raj etc. Andhra Pradesh is the largest producer of brinjal followed by Maharashtra, Karnataka and Orissa. The other main states growing brinjal are Madhya Pradesh,

West Bengal, Uttar Pradesh, Tamil Nadu, Assam, Rajasthan, Bihar, Gujarat, Punjab and Haryana etc.

The cultivars of low glycoalkaloids content with different sizes, shapes and colour increased the market acceptability of the fruits. Resistance breeding for bacterial wilt sustained the cultivation of brinjal in sick soils. The development of Bt-transgenics will lower shoot and fruit borer damage and ultimately the use of pesticides. The standardization of cultural practices, irrigation and nutritional requirement of different cultivars under different soil and climatic conditions help in better crop stand. The availability of brinjal during off-season is now possible when grown under protected cultivation. Use of growth regulators induce parthenocarpy, earliness and yield while different chemical formations help in checking of various insect-pest and diseases of brinjal. Grafting of brinjal cultivars on perennial and wild species increased the yield and availability period of the fruit.

Chapter 2
Nutritive Value and Uses

Brinjal has been a stable vegetable in our diet since ancient times. It is popular, tasty and quite high in nutritive value, used in a number of dishes. It is grown mainly for its green, tender and immature fruits, which are used as a vegetable but sometimes the fruits are dried and cooked with other vegetables. The nutritive value of brinjal is given in Table 6. The unripe tender and succulent brinjal fruits are cooked in a variety of ways. It may be primarily cooked separately or along with other vegetables *viz.* potato, peas, kale, orach, spinach, little mallow, and Nadroo (*Nelumbo nucifera* Gaerth). It makes a special preparation if cooked with moong dal. In some regions of our country like Kashmir the fruit is given 2–3 longitudinal cuts after its harvest, but without separating the pieces and then the sliced fruit is sun dried and sun dried stuff is used during winter months. This practice of drying brinjal is in vogue since time immemorial. It has great potential as a raw material for pickle making and dehydration industries for vast domestic market as well as for export. The discolouration in brinjal fruits is attributed to high polyphenol oxidase activity. The skin of brinjal contains *delphinidin 3-rutinoside* plus a little *delphinidin 3-rutinoside 5-glucoside* (Tanckev *et al.*, 1970). The cultivars which are least susceptible to discolouration are considered suitable for processing purpose. Sidhu *et al.* (1982) in their studies, concluded that there is much variation in the chemical constituents in fruits of different cultivars and chlorophyll, true protein and total phenols are influenced by other constituents of the fruit besides the effect of dry matter, especially on anthocyanin and orthodihydroxy phenols. As a native plant it is widely used in Indian cuisine for example in samber, chutney, curries and achaar and in other Indian foods during festivals. It is often described as the "King of vegetables". In one dish, brinjal is stuffed with ground coconut, peanuts and masala and then cooked in oil. Once prepared, brinjal may be marinated, stuffed, grilled, fried, baked or stir–fried. It is an essential ingredient in several savoury dishes, most notably *baingan ka bhartha*.

Chemical Constituents

The fruits contain solasonine, solasodine, diosgenin, yamogenin, β-sitoserol, arginine, solanine, aspartic acid, histidine, leucine, methionine, pipecolic acid, phenylalanine, theonine, tryptophan, valine, choline, nicotinic acid, riboflavin, vitamin A and C, fructose, glucose, sucrose, anthocyanin, lycoxanthin, caffeic acid, chlorogenic acid and aspartic acid.

The seeds contain steroidal saponins as melangosides-A,B,E,F,H,K,L,M,N,O and P; along with tigogenin, diosgenin γ-hydroxyglutanic acid, lanost-8-en-3-β-ol, lansoterol, 24-methylene lanot-8-en-3-β-ol, cycloartenol, cycloartanol, 24-methylcycloartanol, lupeol, β-amyrin, daturaolone, daturadiol and melongosides N.O and P.

The nutritive value of brinjal can well be compared with tomato as shown in the Table 6.

Health and Nutritional Benefits of Eating Brinjal

Uses

☆ Brinjal fruit is of a high nutritive value and contains vitamin A and B being low in calorie content it is suitable for people who are suffering from high blood pressure, diabetes and obesity.

☆ Although it has low caloric content, but when fried its calorific value rises steeply.

Table 6: Chemical composition of Brinjal Fruits per 100g of Edible Portion

Moisture (g)	92.7	Carotene (µg)	74.0
Protein (Nx6.25)	1.4	Thiamine (mg)	0.04
Fat (g)	0.3	Riboflavin (mg)	0.11
Minerals (g)	0.3	Niacin (mg)	0.9
Fibre (g)	1.3	Vitamin C (mg)	12.0
Carbohydrates (g)	4.0	Choline	52.0
Oxalic acid (mg)	18.0	β-carotene (µg)	0.74
Energy (Kcal)	24.0	Potassium (mg)	2.00
Calcium (mg)	18.0	Copper (mg)	0.17
Phosphorus (mg)	47.0	Manganese (mg)	0.13
Iron (mg)	0.38	Zinc (mg)	0.22
Magnesium (mg)	16.0	Chromium (mg)	0.007
Sodium (mg)	3.0	Sulphur (mg)	44.0
Phosphorus (mg)	47.0	Iron (mg)	0.90

Contd...

Table 6–Contd...

Approximate Total N (g/100gms)	Argi-nine	Histi-dine	Lysine	Trypto-phan	Phenyl-alanine	Tyro-sine	Mithio-nine	Cys-tine	Threo-nine	Leu-cine	Isoleu-cine	Valine
						mg.per gm N						
0.22	210	130	330	060	250	240	070	030	230	380	270	320

Oxalic Acid	Phytin P	Phytin P as per cent of P	Total Dietary Fibre
	mg/100gms		
(1)	(2)	(3)	(4)
18	3	6	—

Source: Gopalan et al., 1995.

Nutritioal Value. 100g (3.50z)

Energy 20Kcal	100kj	Vitamin B6	0.084mg	6 per cent
Carbohydrates	5.7g	Folate (vit. B9)	22µg	6 per cent
Sugars	2.35g	Vitamin-C	2.2mg	4 per cent
Diatary fibre	3.4g	Calcium	09mg	1 per cent
Fat	0.19g	Iron	0.24mg	2 per cent
Protein	1.01g	Magnesium	14mg	4 per cent
Thiamine (vit.B1)	0.039mg	Phosphorus	25mg	4 per cent
Riboflavin (vit.B2)	0.037mg	Zinc	0.16mg	2 per cent
Niacin (vit.B3)	0.646mg	Potassium	230mg	5 per cent
Pantothenic acid (B5)	0.281mg			
Thiamine (vit.B1)	3 per cent			
Riboflavin (vit.B2)	2 per cent			
Niacin (vit.B3)	4 per cent			
Pantothenic acid (B5)	6 per cent			

☆ It is consumed in a number of ways like grilling, roasting, barbeuing, frying and is prepared alone or in combination with other vegetables.

☆ The immature fruits are primarily used as a cooking vegetable.

☆ Besides their use as vegetable, they are used to produce chutneys and pickles and in processing industries.

☆ The fruit is made into 'bharta', a preparation relished in most part of our country, by roasting, mashing and seasoning with salt, onion, chillies, tomato, coriander leaves and fatty oil. The fruit is also sliced and fried or boiled.

☆ In some parts of the country it is sliced and sun dried for off season.

☆ The tender fruits are sometimes used as curries.

☆ It contains medicinal properties and white brinjal is said to be beneficial to diabetic patients.

☆ Field brinjal fruit in til oil can cure toothache.

☆ The roots are useful in asthama, cardiac debility, inflammations, neuralgia and ulcers in the nose.

☆ The root paste is applied on goiter, mumps and hydrocele.

☆ The leaves are useful in asthama, bronchitis, cholera, dysuria, fever and odontolgia.

☆ The leaf juice 10 gm added with 50ml milk and sugar brings sound sleep.

☆ The ripe fruit is digestive, removes gasses and colic, increases appetite and brings sound sleep.

☆ The unripe fruits are useful in sterility, to increase iron in blood, in rheumatism, liver diseases and cardiac debility.

☆ The seeds are useful for decreasing serum cholesterol, in piles and haemorrhoids.

Nutritional Profile and Medicinal Value

Untill 18th century brinjal was looked upon in Europe as something nefarious, capable of inducing fever, or epileptic fits. It was even called *Solanun insanum* by the great botanist and taxonomist Linnaeus before he changed. But today the brinjal, is not eaten plain nor used in infusions. It can be cooked in various ways to provide medicinal properties with out resorting to the rich and heavy method of cooking it in oil.

Brinjal is used mainly as a food crop, but it does also have various medicinal uses that make it a valuable addition to the diet. There are many health benefits of brinjal, like it is believed to be a cholesterol regulator and anti diabetic. In particular the fruit helps to lower blood cholesterol levels and is suitable as part of a diet to help regulate high blood pressure. The fruit is anti-haemorrhoidal and hypotensive. It is also used as antidote to poisonous mushrooms. It is bruised with vinegar and used as a poultice for cracked nipples, abscesses and haemorrhoids. The leaves are narcotic. A decoction is applied to discharging sores and internal haemorrhages. A soothing and emollient poultice for the treatment of burns, abscesses, cold sores and similar conditions can be made from the leaves. Its leaves are toxic and should only be used externally. The ashes of the penduncle are used in the treatment of intestinal haemorrhages, piles and toothache. A decoction of the root is astringent.

Brinjals are of different shapes, sizes, colours, tastes and textures. Bitterness in them is due to the presence of glycoalkaloids which vary from 0.37 to 4.83 per cent per 100 g of fresh weight. Purple coloured fruits have high copper content and polyphenol oxidase activity, whereas it is lower in white cultivars. Similarly Fe and catalase were found highest in green cultivars and lowest in white cultivars. The green cultivars have better processing properties than the purple, but the white cultivars lack anthocyanins pigment responsible for imparting the purple colour to the fruit. Nandkarni (1927) has given different medicinal uses of brinjal viz. when pierced allover with a needle and fried in oil, the fruit is employed as a cure for toothache. It has been recommended as an excellent remedy for those suffering from liver complaints. Its green leaves are a source of the appetizer, aphordisiac, cardiotonic and beneficial in 'Vata and Kaph' etc. Whereas, in Unani system of medicine, its roots are used to alleviate pain. Brinjal fruit is used as cardiotonic, laxative and reliever of inflammation. It has cholestrolizing property primarily due to the presence of poly-un-saturated fatty acids (Linoleic and Lenolenic) present in its flesh and seeds (65.1 per cent). The de-cholestrolizing action is mainly due to the presence of magnesium and potassium in its fruit. Brinjal is effective in the treatment of high blood cholesterol hyper-cholesterolemia. It can block the formation of free radicals, help control cholesterol levels and is also a source of folic acid and potassium. Brinjal is known to have some ayurvedic medicinal properties and is said to be good for diabetic patients (Shukla and Naik, 1993). It helps to block the formation of free radicals and is also a source of folic acid and potassium.

It is richer in nicotine than any other edible plant, with a concentration of 100mg/g or 0.01mg/100mg. However, the amount of nicotine from brinjal or any other food is negligible compared to passive smoking. On average, 9kg of brinjal contains about the same amount of nicotine as a cigarette.

In addition to featuring a host of vitamins and minerals, brinjal also contains important phyto-nutrients, many of which have anti-oxidant activity. Phyto-nutrients contained in brinjal include phenolic compounds, such as caffeic acid and chlorogenic acid and flavonoids such as *nasunin*.

Research on brinjal has focused on an anthocyanin phyto-nutrient found in brinjal skin called *nasunin*. *Nasunin* is a potent antioxidant and free radical scavenger that has been shown to protect cell membranes from damage. In animal studies, nasunin has been found to protect the lipids (fats) in brain cell membranes. Cell membranes are almost entirely composed of lipids and are responsible for protecting the cell from free radicals, letting nutrients in and wastes out and receiving instructions from messenger molecules that tell the cell which activities it should perform.

Phenolic compound that function as anti-oxidant in plants are formed to protect themselves against oxidative stress from exposure to the elements, as well as from infection by fungi and bacteria. The predominant phenolic compound found in brinjal cultivars is chlorogenic acid, which is one of the most potent free radical scavenger found in plant tissue. Benefits attributed to chlorogenic acid include antimutagenic (anti cancer), anti-microbial, anti-

LDL (bad cholesterol) and anti-viral activities. In some African countries it is used to treat epilepsy and convulsions. In South East Asia, it is used to treat stomach cancers and measles. Brinjal is sweet, sharp, hot, cures fever, excessive humor of phlegm and fluctuation. It is an appetizer, increases virility and is light. Brinjals size is very delicate and it cures excessive phlegm and bile in the body.

Cholesterol

Brinjals are effective in curing high blood cholesterol, hyper-cholesterol-lemia. It is rich in antioxidant monoterpenes and helps block the formation of free radicals to maintain cholesterol.

Cancer

The purple pigments in brinjals contain phenolic flavonoid phyto-chemicals (called anthocyanin) that have potential health effects against cancer and other problems like aging, inflammation and neurological disorders.

Flatulence

Roasted brinjal soup taken with asafeotida and garlic is very good for getting rid off gas and congestion within the body.

Hypertension

The presence of potassium, an important intera-cellular electrolyte, helps to counteract the hypertension effects of sodium.

Metabolism

Brinjals contain good quantity of vitamins B-complex (B1, B3, B5 and B6). These vitamins are essential for fat, protein and carbohydrate metabolism.

Dietary Fibre

This vegetable is extremely low in fat and calories and is rich in soluble fibre content. 100 gram bringals contain just 24 calories and 9 per cent of RDA of fibre.

Digestion

Meshed brinjal and tomato soup taken with salt and pepper seasoning helps promote digestion and enhance appetite.

Insomnia

If you have problem getting sound sleep, eat baked brinjal with honey daily before bed time. It will cure your insomnia after sometime.

Enlarged Spleen

It is a valuable remedy for treating enlarged spleen caused by malaria. Consume soft baked brinjal with raw sugar on empty stomach for curing it.

Stones

Eating brinjals helps in destroying stones in their early stages.

Wind humor and phlegm can be cured by taking roasted, peeled brinjals with a dash of salt.

Biological Activities

- ☆ The roots are analgesic, cardiotonic, laxative and stimulant.
- ☆ The leaves are antiherpetic, analgesic, narcotic and sialagogue.
- ☆ The unripe fruits are acrid, aphrodisiac, bitter, cardiotonic, haematinic, stomachic and sweet.

☆ The seeds are anticholesterolemic and stimulant.

Toxicity

☆ It was belief that ingestion of brinjal could cause madness.

☆ An early authority warned that brinjal fruit 'endanger meloncholly-the leprosy, cancer, piles, headache, stinking breadth, breed obstructions in the liver, spleen and charge the completion into foul black and yellow colour, unless they be boiled in vinegar".

☆ Regular intake of brinjal is said to cause skin irritation.

☆ The skins of vegetables like brinjal, tomato, bitter gourd and squash are thought to shield them from excessively sprayed chemical. Soaking them in water for 30 minutes will help in over coming its ill effects, as pesticides are mostly water-soluble.

Chapter 3
Taxonomy

Brinjal is a member of the potato family and it is known world wide as aubergine, egg plant, melanzana, garden egg, patlican, berenjena, egg apple, melongene, melanzane and Guinea squash. It is available year–round with peak season during the months of August to September.

A. Vernacular Names

There are several names by which the crop is known in India and abroad, but brinjal is the most familiar throughout the country.

- ☆ *Assamese*: Bengana
- ☆ *Bengali*: Begun
- ☆ *English*: Egg plant, Brinjal, Guinea Squash.
- ☆ *French*: Aubergive

☆ *Gujarati*: Vengan, Ringana

☆ *Hindi*: Baingan

☆ *Kashmiri*: Waangun

☆ *Konkan*: Vayingana

☆ *Kanad*: Badanekaya, Doddabadane

☆ *Marathi*: Vange, Bhata,Vanga, Vangi

☆ *Malayalum*: Vazhuthana

☆ *Mayanmar*: Kayan

☆ *Malaysia*: Terong

☆ *Nepali*: Bhenta

☆ *Oriya*: Baigan

☆ *Punjabi*: Batavan, Bengan

☆ *Persian*: Badangan

☆ *Sanskrit*: Peetaphala, Vartaku, Karttaki, Bhantaki

☆ *Telugu*: Venkaya

☆ *Tamil*: Katharikai

☆ *Thailand*: Makeu-a-Kaou

☆ *Chinese*: Qie zi/Ai Kwa

☆ *German*: Eirfrucht/Aubergine

☆ *Italian*: Melanzana/petonciano

☆ *Japanese*: Nasu

☆ *Russian*: Baklazan

☆ *Spanish*: Berenjena

☆ *West Indies*: Berenjena

☆ *Netherlands*: Eierplant

B. Systematic Position

Kingdom	Plantae
Class	Magnoliopsida
Subclass	Asteridae
Order	Solanales
Family	Solanaceae
Genus	Solanum
Edible species	*S. melongena*
Binomial name	*Solanum melongena*
Chromosome No.	24 (36,48)

Synonyms

☆ *Solanum ovigerun* Dunal

☆ *Solanum trongum* Poir

Genus Solanum comprises approximately 2000 species in 70 genera, which include both tuber bearing and non tuber bearing forms, of which about 150–200 are tube bearing and belongs to section *Tuberarium* and majority of species are (about 1,800) are non-tuber bearing and the edible belongs to this section. A number of these are important medicinal and food plants. One such genera is Solanum which holds some 900 species with numerous economically important crop plants. These are commonly called as brinjal, egg plant and guinea squash are some synonyms to it.

The important edible species under non-tuber bearing forms are *S. melongena, S. torvum, S. nigrum* and *S. macrocarpon., S. ferox* and *S. aethiopicum.*

The species *S. anomalum, S. macrocarpon* L., *S. aethiopicum* L., *S. incanum* L., *S. duplosinuatum* and several other minor species are cultivated in Africa. The species *S. quitoense* Lam., *S. muricatum* Ait. and *S. topiro* Humb and Bonpl. are cultivated in Central and South America. The species *S. aethiopicum* L. and *S. macrocarpon* L. are cultivated not only for fruits but also for their edible leaves. In South-East Asia small fruits of *S. ferox* L. and *S. Torvum* Sw. are eaten.

☆ *S. torvum*–It has small clustered fruits which are used for curry purposes and for drying. It is grown as a wild plant in backyards and road sides. Due to its resistance to fusarium wilt and bacterial wilt, it can be a resistant root stock for grafting cultivated *S. melongena.*

☆ *S. macrocarpon* and *S. aethiopicum* are generally grown for their edible fruits and leaves.

☆ *S. nigrum*–It has small clustered acidic edible fruits which are ready for harvest when purple in colour. In Tamil Nadu *S. nigrum* is cultivated for its edible leaves. The ripe fruits are also used in curries and for frying. It leaves and fruits resemble to that of chilli.

Several of Solanum species for example *S. melongena* var. *insanum*, *S. incanum*, *S. indicum*, *S. viarum* are locally used as medicine throughout the tropics.

Being indigenous to India, a lot of variability is available. Many centres have collected and maintained the wild species of Solanum such as *S. melongena* var. *insanum*,

S. torvum, S. nigrum, S. incanum, S. indicum, S. khasianum, S. surratance, S. trilobatum, S. mamosum, S. sisymbrifolium and *S. gigantium.*

Studies on inter-specific hybridization revealed incompatibility between cultivated *S. melongena* with other species, except with *S. insanum.* Further taxonomic studies revealed that *S. insanum* is only a variety under *S. melongena* and can be treated as *S.melongena* var. *insanum.*

S. melongena var. *melongena (Syn: melongena* var. *esculenta* Nees) includes cultivars with round, long and egg shaped fruits.

S. melongena var. *serpentinum* Desf. includes long and slender fruited cultivars (snake brinjal, fruit is extra ordinarily long, 2.0 cm or less in diameter and 30-40 cm long).

S. melongena var. *depressum* L. includes early and dwarf cultivars (plant is extremely short and dwarf) produce small pear-shaped fruits that are purple in colour.

S. melongena var. *insanum* includes wild and prickly plants with small fruits.

S. melongena var. *bulsarensis* L.

Classification on the Basis of Fruit Shape and Plant Spread

These are: Long brinjal (Pusa Purple Long, Pusa purple cluster).

Round brinjal (Pusa Purple Round, T-3).

Oval brinjal (Pusa Kranti).

Classification on the Basis of Fruit Colour

These are purple brinjal (anthocyanin present) such as Pusa Purple long and green brinjal (chlorophyll present) *e.g.* Arka Kusumkar.

Solanum which is a very large genus. Among the 22 Indian species, there is a group of 5 related ones, all prickly and diploids, namely *melongena, coagulans, xanthocarpum, indicum* and *maccanii* (Choudhury, 1976). Out of these species, *solanum incanum* (often considered synonymous with *S. coagulans*) produces fertile hybrids in crosses with *S. melongena*. It appears that *melongena is* more closely related to *S. incanum* than to any other species. Occurrence of natural hybridization between *S. incanum* and *S. melongena* has been reported by Viswanathan (1975). Two other species, *S. khasianum* and *S. aviculare*, having same chromosome number, have gained importance as source of *Solasodine* used for synthesis of steroid hormone (Choudhary, 1976). The cross between *S. khasianum* and *S. melongena* is only successful when *S. melongena* is used as a female parent.

Botany

It is a tender perennial grown as an annual and is characterized by a bush, indeterminate erect, profusely branched herbaceous plant attaining a height of 0.6 to 1.5m. The leaves are large, lobed, hirsute on the under side, occasionally bearing sharp spines and variable in habit. Brinjal is usually self-pollinated but the extent of cross pollination has been reported as high as 29 per cent and hence it is classified as *often cross pollinated* or facultative cross pollinator. Inflorescence is often solitary but sometimes it constitutes a cluster of 2–5 flowers. Solitary or clustering nature of inflorescence is a varietal character.

- ☆ It is a bushy plant that grows erect to about 60–120 cm tall.
- ☆ Its growth is indeterminate.
- ☆ The fruit is berry, borne singly or in clusters.
- ☆ It is a self pollinated crop, but with cross pollination to the extent of 30 to 40 per cent.
- ☆ Four types of flowers are their in brinjal.
- ☆ They are long styled and medium styled flowers producing fruits whereas pseudoshort and true short styled flowers do not set any fruit.
- ☆ The fruit setting in brinjal varieties with long styled flowers varies from 70 to 86 per cent, while short styled flowers do not set fruit.
- ☆ The fruit is a pendent, fleshy berry that can be oval, round or oblong, 5 cm to 20 cm in diameter.
- ☆ The epidermis or skin of the fruit is thin and smooth.
- ☆ The fruit is mostly fleshy placenta, in which the seeds are embedded in a mass of spongy tissue.
- ☆ Most cultivars have purple fruits, although colours may range from yellowish white to red and to blackish purple.

Brinjal is a diploid with $2n=2x=24$. Plant is erect, semi erect, compact or prostrate, polymorphous, herbaceous, woody at the base and branched with around 0.5–1.5m height. The crown is broad, branches are light green or tinged with violet, rather thick, densely clothed with gray, stellate tomentum.

Roots

Vigorous, deep growing tap root. The crown is broad. Branches are light green or tinged with violet, rather thick, densely clothed with grey and stellate tomentum

Stem

The *stem* is spiny or non-spiny with or without purple pigmentation which is due to anthocyanin. Hairy, woody at base, some times with violet tinge.

Leaves

Leaves are alternate, solitary or in pairs, on shorter or longer stalks, large, simple, ovate or oblong-ovate-elliptic. Base is often unequal sided and oblique, rounded, truncate or cordate, apex triangular, acuminate, obtuse or acute; irregularly undulate, angular or more or less deeply lobed and above dark green. On both sides dull, densely clothed with soft, gray, stellate tomentum, flaccid, 7–25 cm long, 5–15 cm wide, reticulate- veined and veins transparent. The underside of most cultivars is covered with dense wool-like hairs.

Petiole

Petiole is robust, subterete, yellowish green, deeply white stellate tomentose and 1.5–1.0 cm long. Lamina ovate, sometimes lobed, hairy, pointed at apex, margins undulate.

Flowers

Flowers are bisexual, pentamerous, and are rather large, stalked, lateral or leaf-opposed, the female ones solitary, the males in many-flowered inflorescence. Flower is complete, actinomorphic and hermaphrodite. Pedicel 1-2 cm long, is light green or strongly tinged with violet, terete, aculeate or not, densely stellate tomentose, patent or curved and 1.5–2 cm long. The lavender flowers, borne singly or in small clusters are similar to tomato, but larger.

Calyx

Calyx is persistent and spiny or non-spiny, tubular-campanulate, 5–lobed far less than half way down, on the outside grayish green, often strongly tinged with purple, on both sides densely stellate- tomentose and 1.5–2 cm long, 5-6 lobed. The calyx is gamosepalous and persistant. It forms a cup like structure at the base. Tube is ovoid- oblong, obtusely pentagonous, ± 1 cm long, 0.5–0.7 cm diameter; segments narrowly oblong, acuminate, obtuse, obliquely erect and 0.5–1.5 cm long.

Corolla

Corolla is gamopetalous, deeply 5-lobed, stellately spreading, on the outside light violet, densely white stellate tomentose, within dark violet, glabrous, rugose and 2.5–4 cm in diameter. Lobes are short, ovate-oblong, and rather acute.

Stamens

Stamens are 5 erect, ovoid-oblong, vitelline, opening at the top by two pores and 0.5 cm long. Dehiscence is poricidal. Anthers are cone shaped, free and with apical dehiscence.

Ovary

Ovary is bilocular with many ovules, hypogynous, bicarpellary, syncarpous and with basal placentation. Heterostyle is very common.

Berries

The fruits are berries, are most variable as to shape, colour and dimensions, globose or oblong, white, yellow or more or less intensely violet, glabrous, 2.5–15 cm long and 2.5–4 cm in diameter, smooth and shiny.

Fruiting

Fruiting calyx is campanulate, much enlarged, bursting on one or two sides, pale green or tinged with purple, thinly stellate tomentose, unarmed or beset with scattered prickles, 3–6 cm long and 3–5 cm in diameter. Fruiting pedicel is much elongated, much thickened towards the apex, unarmed or with scattered small prickles, pale green, nodding and 3–7 cm long. Fruit colour may be nearly black, different shapes of purple, white, green or variegated. Fruit setting flowers consist of long and medium styled flowers. Fruit setting in long styled flowers normally varies from 70–85 per cent and that in medium styled flowers from 12 to 55 per cent. The non-fruit setting flowers consist of short styled flowers in which androecium is fertile but stigma is smaller with underdeveloped papillae.

Seeds

Seeds are borne on the fleshy placenta, which together completely fill the cavity (Som and Mallik 1986). Seeds are numerous, small, light brown, approximately 250 seeds/g. Brinjal seeds resemble with those of chillies in colour, shape and appearance. But they are comparatively smaller than chillies seeds.

Though brinjal is *self-pollinated* crop, but there is high degree of cross-pollination (as high as 29 per cent) due to *heteromorphic* flower structure (called as heterostyle) and is mainly by honey bees and bumble bees. They are scattered through the fruit, imbedded in a fruit placenta.

   Depending upon length of the style in relation to position of anthers, four types of flowers–heterostyle are available.

1. Long styled-stigma well above the anthers, fruit set ranges from 60–70 per cent.

2. Medium styled-stigma and anthers at the same level, fruit set ranges between 12.5 to 55.6 per cent.

3. Short styled-style short.

4. Pseudo short styled-style rudimentary.

   Both short styled and pseudo styled flowers act as male flowers therefore there is no fruit formation. However, with application of growth regulators (2,4-D and NOA) even pseudo short-styled flowers show fruit setting up to 60–70 per cent. The fruit setting in brinjal is increased after soaking brinjal seeds in 5 per cent 2,4-D solution for 24 hours before

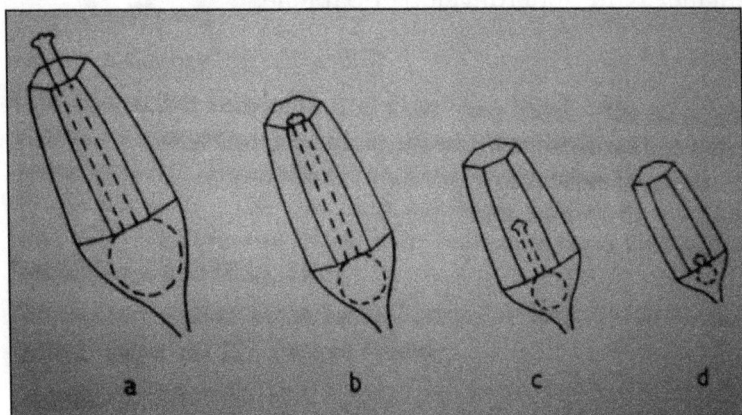

Types of Brinjal Flowers on the Basis of Length of Styles

(a) Long styled (big ovary); (b) Medium styled (medium ovary); (c) Pseudo short styled (rudimentary ovary); (d) True-short styled (very rudimentary ovary)

sowing or spraying the crop with 2 ppm 2,4-D just at the commencement of flowering. Considerable yield improvement is obtained with the application of GA3+ ascorbic acid each at 250 ppm as root dipping of seedlings before transplanting.

Flower Biology and Pollination

Anthesis and Anther Dehiscence

Flowers generally emerge 40–45 days after transplanting, and open mainly in morning. Full bloom is observed 80 days after planting. The whole period of

Brinjal Flower

Brinjal (Fruit)

effective flowering lasts for 75 days. Anthesis in brinjal flower normally starts from 5.35 AM and continues up to 7.35 PM with peak at 6.05 AM in August-September and usually between 9.30 to 11.15 AM during winter (December-January) whereas, the pollen dehiscence begins 30 minutes after anthesis. It commences at 6 AM and continues up to 8.00 AM with the maximum at 6.35 AM. Anther dehisce usually 15-20 minutes after flower bud had opened. Both anthesis and dehiscence depends upon the (Table 7) day light, temperature, cultivar and humidity, however, the period of effective receptivity ranges from a day prior to flower opening to 4 days after opening. Stigma receptivity is highest during anthesis *i.e.* flower opening. Anthers usually dehisce 15-20 minutes after the flowers have opened. The receptivity of stigma could be observed from the pump and sticky appearance, which gradually turn brown with the loss of receptivity. Pollen is most fertile immediately after the anther dehiscence. Pollen usually remains viable for a day during summer and 2-3 days in winter under field conditions. However, pollen grains can remain viable from the day of anthesis to 10 days at the atmospheric temperature of 24.6°C and relative humidity of 82 per cent. Opening of anthers is mostly by a pore or slit at or near the apex. According to Deshpande *et al.* (1978) maximum anthesis occurred at 6 AM in Arka Shirish and at 8 AM in Arka Sheel and Kusumakar. In addition to above factors, soil fertility may also influence flower initiation and development.

☆ The flower of brinjal (*Solanum melongena*) is hermaphroditic, self pollinating and hangs from the plant.

☆ The flower does not produce nectar.

☆ The stamens form a loose ring around the pistil and have an opening at the ends. The stigma usually protrudes just beyond this ring.

☆ In addition to self pollination, cross pollination may occur.

☆ Just a slight movement of the flower is sufficient for the pollen from the stamens to fall onto the stigma.

☆ Bumble bees cause this movement by hanging upside down on the flower, fastening their jaws onto a stamen and then setting the flower into vibration by activating their flight muscles, without making flight movements with their wings (this is called buzz pollination).

☆ In fine flowered varieties, the jaw marks can be recognized by tissue discolouration. Jaw marks indicate that the flower has been visited and therefore, has been "set."

☆ Because brinjal flower does not produce nectar, the hive is supplied with sufficient sugar water from the expected life span of the bumble bee population.

Pollination

Pollination is transfer of pollen from stamens to the pistil. Pollen is transferred primarily by wind (for instances in grasses and conifers) and by insects (many bees, butter flies and moths, in most flowering plants), but also by water and vertebrates such as mice, bats and birds (*e.g.* the humming bird). Pollination can occur both within the same flower and between different flowers, either or not on the same plant. In which of these cases pollination leads to fertilization depends on the specific properties of the plant species. In humid climate cross pollination may occur up to 20 per cent mainly due to heterostyle even though it is treated as a self pollinated crop.

For emasculation, a healthy long or medium styled, well developed bud from the central portion of the plant is selected. The bud is opened gently with the help of fine pointed forceps one or two days before the opening of the bud and all the five anthers are carefully removed. For pollination, freshly dehiscing anthers are picked up and are silt vertically with fine needle to get sufficient pollen flower bud. It is labeled and covered with small pollination bag.

Table 7: Floral Biology of Brinjal

Anthesis	Dehiscence	Pollen Fertility	Stigma Receptivity	Ref.
In the morning hours	Just after anthesis	For about 2 days	On the day of anthesis to 2 days after anthesis	Kakizaki, Y., 1924, 1930
6.00–11.30 hr	4.15–11.00hr on the next morning also	On the day of anthesis to 2 days after anthesis	2–3hr before anthesis to 2 days after anthesis	Pal and Singh, 1943
In the morning hr.	–	On the day of anthesis to 7–10 days after anthesis	One day before and 6-8 days after anthesis	Popova, 1958
7.00–14.00hr	7.15–14.00hr	On the day of anthesis	One day before to 2 days after anthesis	Prasad and Prakash, 1968

Chromosome Number

The cytological studies revealed that *Solanum melongena* contains chromosome number $2n=24$ (normal diploid). Triploids with $2n=36$, and tetraploid with $2n=48$ have been produced with non-significant economic use (Chandrasekaran and Parthasarthy, 1948).

Hybridization

At I.C. A.R New Delhi a high yielding cultivar Pusa Kranti was developed for northern plains of India. The cultivar was bred by combining good qualities of three parents *viz.* Pusa Purple Long, Hyderpur and Wynad Giant. The resultant hybrid (Pusa Kranti) possesses photo-insensitive character from Pusa Purple Long, up right habit, hardiness and glossiness from Hyderpur and the bright purple colour of Wynad Giant. Pusa Purple Long was first crossed with Hyderpur and the progeny so developed was back crossed to Wynad, hybrid pusa kranti was thus evolved through selection from subsequent generations.

In hybridization programe emasculation forms a major bulk of work. Jyotishi and Chandra (1969) found that foliar sprays with 10ppm of 2,4-D at 10 days interval resulted in complete pollen sterility in the egg plants cv. Pusa Purple Long. Open cross pollination from male parent plants grown in alternate rows the flowers from treated female plants gave a fair fruit set and an average of 109 seeds per fruit. Although the number of seeds per fruit obtained by this method is much smaller than obtained by hand emasculation and hand pollination.

Some wild species were also used in the hybridization programmes of egg plants with a view to bring some characters such as disease resistance, pest and drought resistance etc. Gopimony and Sreenivasan (1991) observed that the F1 hybrids of three egg plant cultivars crossed with wild *Solanum melongena* var. incanum showed a high degree of heterosis in branch, flower and fruit numbers and have much longer tap root.

The occurrence of natural hybridization between *S. incanum* and *S. melongena* L. was demonstrated by Viswanathan (1975). Seedlings were raised from highly spiny form of *S. incanum* found growing wild. One aberrant seedling had large flowers and purple fruits. Self seedlings from this aberrant plant had fruits which segregated for size and colour the largest resembling a round *S. melongena* and smallest being similar to *S. incanum*.

Hybrid Vigour

Hybrid vigour in brinjal was recorded as early as 1892 by Munson in U.S.A and by Nagai and Kida during 1926 in Japan. Today it is a common practice in Japan to use brinjal hybrids.

In India Pal and Singh (1946, 1947) reported that hybrid brinjal increased yield of 48.8 to 56.6 per cent over the better parent. Pusa Purple a best hybrid with uniform and attractive fruits was evolved from a cross between Clustered White and Muktakeshi strains. In 1961 Mishra observed increased yield in brinjal hybrids. While as Viswanathan (1973) reported 100 per cent increased yield in a cross between Muktakeshi x Banaras Giant over the parental mean. In West Bengal Som and Malik (1979) observed markedly higher yield of about 164 and 134.10 per cent in a cross between Kalyani Green and Elokeshiover due to increase in the number of branches, fruit size, fruit number and fruit weight.

Pusa Anmol a cross between Pusa Purple Long and Hyderpur, evolved at I.C.A.R New Delhi, is an early and recorded an increased yield of about 80 to 100 per cent over PPL (Choudhary, 1966). Thakur *et al.* (1968), Lal *et al.*(1974) and Singh *et al.* (1978), also reported hybrid vigour in brinjal.

Exploitation of Heterosis

The exploitation of heterosis in brinjal for fruit yield, quality and resistance to diseases has received the attention of the breeders due to hardy nature of the crop, comparatively large size of flowers, and the large number of seeds in a single fruit, thus enabling production of a large seeds with a single act of pollination, as hand emasculation and pollination are still followed in hybrid seed production in brinjal. Consumer acceptance which varies from region to region demands development of large number of high yielding F1 hybrids. F1 hybrids are becoming popular day by day where manual emasculation and pollination are very much in practical use. Quite a large number of heterotic hybrids have been developed and are being developed at ICAR-Institutions and State agricultural Universities in the country. In addition, a large number of F1 hybrids are marketed by private seed companies *e.g.* Supriya, Sourabha, Suphal (IAHS), Kalpatharu, Ravaiya (Mahyco), Kanhaiya, Novkiran, Pragathi (Sungro Seeds), Aspara, Nisha (Namdhari) etc. are a few commercial hybrids popular among farmers. Some important ones are detailed in Table 8.

Genetics

Very few studies have been made on genetics of *qualitative* characters in brinjal.

Table 8

Developing Institute	Hybrid	Parents	Special features
IARI, New Delhi	Pusa Anmol	Pusa purple long × Hyderpur	Produce 80 per cent more than PPL. Yield 62t/ha
Pusa Hybrid-5		—	Long glossy dark purple fruits. Yield 51.6t/ha.
Pusa Hybrid-6		—	Early, round yield 45.0t/ha.
Pusa Hybrid-9		—	Early dark purple round fruits. Yield 56.0t/ha.
IIHR, Bangalore	Arka Navneeth	IIHR 221X Supreme	Large dark purple round to slightly oval fruits. Yield 65-75t/ha.
	Arka Anada	IIHR-3 × SM6-6	Fruits green long, medium sized (50-55g). Yield 65t/ha. Resistant to Bacterial wilt.
Haryana Agri. University	Hissar Shyamal (H8)	Aushey × BR112	Fruits round bright purple, resistant to bacterial wilt, little leaf
Kerala Agri. University	Neelima	Surya × SM116	Round to oval purple fruits. Resistant to bacterial wilt. Protected fruiting. Yield 62.0t/ha
GAU, Anand	ABH-1	M2 × M35	Early, purple oval fruits. Yield 37.0t/ha.
CSAUA&T, Kanpur	Azad Kranti	Pusa purple long × BGL	Long, dark purple fruits.
GBPUA&T, Pant nagar	Pant Brinjal Hybrid-1	PB121 × PB225	Fruits long, borne in clusters. Tolerant to bacterial wilt.

Baha-Eldin and Blackhurst (1968) reported partial dominance of tall plants over short plants, early flowering over late flowering and round fruits over long fruits in a cross between Black Beauty and P1 169–651.

Swamy Rao (1970) observed dominance and partial dominance of long fruit over round fruit. He further observed dominance of clustered fruit over non-clustered fruit.

Rangaswamy and Kadambavanasundaram (1973) reported that a pair of duplicate genes appeared to control brinjal seedling colour.

Thakur *et al.* (1969) reported that fruit colour in brinjal is controlled by dominant complementary gene. They further were of the view that purple corolla is mono-genically dominant over white and green flower and the presence of spike is controlled by a single dominant gene.

According to Sambandam (1964) the purple hypocotyl is dominant over green and controlled by a single gene, he further reported in 1969 that fruit skin and flesh colour are inherited independently of each other and concluded that purple pigmentation in different parts of plant is controlled by several linked genes.

The genes Ph, Al, Pb, Pm, Pu, Pl, St, and C are concerned with the pigmentation of the hypocotyl, epicotyl, leaf midrib, leaf margin, upper and lower surface of calyx, anther suture and fruit skin respectively. The genes PH, Pb, Pu, Pl, St and C are linked.

Gene Studies in Brinjal

In India, the National Bureau of Plant Genetic Research is the nodal institute for management of germplasm resources of crop plants and holds more than 2500 accessions of brinjal in its gene bank. The wide regional variations for plant, flower and fruit descriptions revealed enough scope for improvement of yield characters by selection.

Characteristics	Gene Effects
Plant height	Dominant/Partial dominance
Early flowering	Additive/Dominance/ Partial dominance
Fruit shape index	Partial dominance
Fruit length	Additive/Partial dominance
Fruit girth	Additive
Fruit weight	Additive dominance
Fruit number	Additive/Dominance/ Partial dominance
Number of flowers/ inflorescence	Additive
Branches/plant	Dominance
Seeds/fruit	Dominance
Fruit yield/plant	Additive/Dominance

Kalloo, 1988.

Quantitative

Quantitative characters like yield, size, shape, number of branches and number of fruits are poly-genetically inherited in brinjal even resistance to diseases and pests is sometimes poly-genetically controlled. Studies on plant, stigma, corolla, fruit and flesh colour, presence of strips and spines prove that they are also of the qualitative characters found to be inherited of offsprings (Thakur *et al.*, 1969). Unlike qualitative characters which have a simple inheritance, the polygenic characters are greatly influenced by environment, hence a phenotype is often not a true indicative of its genotype.

A quantitative character, the cumulative effects of the individual genes and their interactions, both allelic or dominance and non-allelic or epistasis, determine the genotypic value.

Peter and Singh (1973) observed additive gene action in respect of number of fruits per plant, fruit length, number of flowers per inflorescence, number of long styled and medium-styled flowers in brinjal. They further reported that maturity and number of primary branches are controlled by over-dominant genes. While as fruit weight per plant is controlled by dominant genes and the height of the plant is controlled by additive gene action with some over-dominance.

While Gill *et al.* (1976) and Gowda (1977) observed in brinjal, that additive gene effects to be more important than dominance in most of the economic characters. Selections in the segregating generations of the crosses are suggested for bringing out the desired improvement in this crop.

Qualitative Genes

The available information based on Chadha (1993) and Kalloo (1993) is summarized in Table 10.

Table 10: Qualitative Genes in Brinjal

Trait	Genetics
Hypocotyl colour	Purple dominant over green, monogenic.
Stem colour	Purple dominant over green, monogenic.
Prickly nature	Prickly nature dominant to non-prickly. Monogenic
Fruit clusterness	Clusterness partially dominant to solitary nature. Monogenic.

Trait	Genetics
Spines	Presence of spines dominant over absence, monogenic
Plant height	Tall dominant over dwarf, monogenic
Fruit colour	Purple dominant over green, green dominant over white, monogenic, digenic, trigenic.
Fruit shape	Elongated dominant over round, round dominant over oval, 3-4 gene control.
Fruit flesh colour	Green dominant over white, monogenic
Bearing habit	Clustering dominant over solitary fruiting, single gene.
Male sterility	ms 1 and ms 2
Style erectness	Incured dominant over straight, monogenic
Bacterial wilt	Single dominant gene for resistance

Breeding Objectives

The main objectives are to evolve hybrids that have following traits:

☆ High yield

☆ Earliness

☆ Fruit shape, size and colour as per the consumers preference

☆ Low proportion of seed and more pulp.

☆ Soft flesh

☆ Low olanine content

☆ Upright sturdy plant free from lodging.

☆ Give flowering and fruiting at very and high temperatures.

☆ Having straight growth habit

☆ Good attractive colour.

☆ Less fruit thickness.

☆ Resistance to:

1. Bacterial wilt (*Pseudomonas solanaceanum*)
2. Phomopsis blight (*Phomopsis vexans*)
3. Little leaf (*Mycoplasma* like bodies)
4. Root knot nematodes
5. Shoot and fruit borer
6. To frost.

Chapter 4

Breeding Methods/ Selection Criteria

The main changes to the fruit which have resulted from many years of plant breeding include a much better tasting fruit, often much bigger fruit, lots of different colours and sizes and also much higher yield *i.e.* more fruit per plant.

For production of resistance breeding, the common breeding methods employed to brinjal are:

Pure Line Selection

Most of the breeding work done in India on brinjal has been through pure line selection from local populations and hybridization. The brinjal plant with stands inbreeding depression and shows a marked response to heterosis. Pure line is the progeny of a single self-fertilized homozygous individual. This procedure may be applied to improve any local variety or to select useful variability existing in indigenous or exotic collections. In case there is variation within plants of a progeny, single plant selection is followed and the same process continues till complete homozygosity is attained. In vegetable crops, pure line improvement is adopted to improve heterogeneous foundation stocks and as a result, varietal improvement is possible. In a self-pollinated species, all the plants are expected to be homozygous because of continued self-fertilization. Thus progeny of a single plant from a population of a self-pollinated crop would be a pure line.

Varieties bred through pure line selection as indicated below.

Pusa Purple Long is a selection from the Mixed Batia cultivar grown in Punjab, Delhi and Western Uttar Pradesh. CO.1 is a pureline selection with crop duration of 160 days. The fruits are oblong and medium sized (50 to 60g) with pale green shade under white background. CO.2 is a pureline selection from the local variety "Varikkathiri" in the village Negamum. The fruits are oblong with dark purple stripes intermingled with light green colour, yielding about 35t/ha in a span of 150 days. MDU.1 is a selection from Kallampatti local type near Madurai, producing record yield of 34 tonnes per hectare in 135 to 145 day of crop duration. KKM.1 is another pure line selection from Kulathur local near Tirunelveli, producing 2-4 egg shaped, milky coloured fruits in clusters in a crop, yielding about 37 tonnes per hectare duration of 130 to 135 days.

Pedigree Method

This method was adopted to develop material as well a variety for high yield, earliness, quality and resistance to pests, diseases and stress conditions. In several crops improvement has been accomplished through pedigree method and as a result, a large number of varieties have been developed. In this method, single plant selection is followed up to F5 or F6 generations and in advance generations, families are selected on the basis of their phenotypic performance.

Many varieties have been developed through hybridization and subsequent pedigree selection. Pusa Kranti has been developed at IARI, New Delhi from the cross PPL × Hyderpur × Wynad Giant.

Bulk Method

This method consists of growing large populations in each generation and harvesting the seeds in bulk and planting a sample of seeds in the following years. No selection is made in F2 and bulk populations grown up to F5 or F6. In this case only natural selection operates. At the end of the bulking period, individual plants are selected and evaluated in a similar manner as in the pedigree method of breeding. The difference between the bulk and the pedigree methods, lies in the manner in which the segregating generations are handled. In pedigree method, individual plant progenies are grown and evaluated in F3 and subsequent generations, but in the bulk method these generations are grown as bulks.

Back Crossing Method

In the back cross method, the hybrid and the progenies in the subsequent generations are repeatedly back crossed to one of the parents. This method involves the selection of two parents where one is the recurrent (agronomic) and the other is the non-recurrent (donor) parent. The F1 population is grown in rows and tested for resistance. This method is best executed in self-fertilized crops than in cross fertilized crops and is usually applied to transfer–resistance characteristics controlled by major genes in cultivars, or to transfer some morphological traits or to improve male sterile lines.

Heterosis Breeding

It is the superiority of F1 hybrid over both the parents in terms of an increased vigour, size, growth rate, yield and a number of other characteristics. For example, many F1 hybrids in tomato are earlier than the parents.

Emasculation is done on buds, which are about to open on the next day morning. Normally, single, long styled flowers are selected for crossing. After emasculation, the flowers are covered with butter paper cover and pollination is done on the next day morning. Generally pollen remains viable for three days. However, viability starts declining from first day onwards. Stigma receptivity and fertilizing capacity of pollen are the highest at the time of opening of floral parts.

Many F1 hybrids have been developed and released for commercial cultivation, *viz.*, Arka Navneet (F1) is a cross between IIHR22-1 × Supreme from IIHR, Bangalore. The fruits are oval, attractive, free from bitterness, flesh soft white with few seeds and deep purple in colour. The yield recorded is 65–70 tonnes per hectare. Another important variety developed through heterosis breeding COBHI–It is a cross between EP45 × CO_2. The fruits are medium sized, oblong and purple deep in colour, with a yield potential of 55–60 tonnes per hectare.

Mutation Breeding

PKM-1, developed at TNAU, Periakulam. Is an induced mutant (gamma rays) of a local type called "Puzhuthi Kathiri" It is a drought tolerant and can with stand long transportation, yielding on an average 34.75 tonnes per hectare in a duration of 150-155 days. The fruit weighs 45- to 65 g.

Resistant Breeding

Brinjal like any other vegetable crop is susceptible to number of diseases, pests and root knot nematode. Therefore, screening of cultivars and species of brinjal for resistance against diseases and pests is of great significance for the evolving of resistant cultivars. Therefore, efforts are in progress at different research stations especially at I.A.R.I, New Delhi for finding sources of resistance against various diseases and pest. Chakraborty and Choudhury (1975) reported selection of three brinjal lines *viz.* Sel.212-1, Sel. 252-1 and Sel.252-1-2 as free from little leaf disease out of 164 germplasms tested. Similarly Kalda *et al.* (1976) found that *S. xanthocarpum, S. indicum, S. gilo, S. khasianum, S. nigram* and *S. sisymbrifolium* are highly resistant to phomopsis blight and *S. melongena* lines 114 and 264 are found resistant. They further stated that resistance to this disease is recessive and polygenically inherited and dominance gene effects are more pronounced than additive. In 1951 Decker released two brinjal cultivars *viz.* Florida Market and Florida Beauty resistant to this blight. During 1969 Srinivasan and Gopimony reported that a wild variety *viz. S. melongena* var. *insanum* is resistant to bacterial wilt (*Pseudomonas solanacearum*). Pusa Purple Cluster is also found to be resistant to bacterial wilt (Rao *et al.*, 1976).

According to Swaminathan and Srinivasan, 1971) *S. melongena* var. *insanum* carries a single dominant gene for resistance wilt and cultivar Black Beauty and Vijaya are tolerant to root-knot nematodes (*M. incognito*) as reported by Yadav *et al.* (1975).

Dhankar *et al.* (1977) found that *S. sisymbrifolium* and PPC-2 are least susceptible lines to shoot and fruit borer. While as, *S. torvum* is reported to be highly resistant to Epliachna beetle.

Fusarium wilt resistance was observed in *S. incanum.* Florida Market, Harris 468 special, Hibush and Harish Hybrid 77631 were highly resistant to *Verticillium dahliae.* Bacterial wilt caused by *P. solanacearum* which has three races *i.e.* 1, 2 and 3 out of which 1 and 3 are more pathogenic than race 2, is one of the most serious diseases of brinjal in warm humid areas. The brinjal lines SM1, SM6, SM48 SM56, SM70, SM72 and SM74 and one line of *S. texanum* were also found resistant to bacterial wilt. PPR is found to be tolerant to little leaf disease of brinjal.

Breeding for resistance to little leaf and shoot and fruit borer are receiving much attention. The resistance however, is not available in the cultivated varieties. Since, certain wild species (*S. incanum, S. anthocarpum. S. khasianam* and *S. sisymbrifolium* are reported to be resistant, they can be used in the hybridization programe. However, the inferior quality of the fruits of wild species associated with resistance are expressed in hybrid progenies.

Cytogenetics and Plant Breeding

Cytological studies revealed that *Solanum melongena* contains normal diploid (2n=24), triploids (2n=36) and tetraploids with 2n=48 chromosomes (Chandrasekaran and Parthasarathy, 1948).

Thakure *et al.* (1969) reported that stigma, corolla, fruit and flesh colour, presence of strips and spines are some of the qualitative characters found to be inherited by off-springs. Likewise, Murtazov *et al.* in 1969 showed that the F1 plants obtained from cross between Patladzan 12 x Delicatesse and Plodiv 43 x Izraelski sown at the beginning of March, April and May showed a tendency towards greater earliness and yield than in February.

In hybridization programmes emasculation is a time consuming process therefore, to overcome it, some chemicals are tried to induce pollen sterility. In this respect Jyotishi and Chandra (1969) reported complete pollen sterility when Pusa Purple Long cultivar of brinjal was sprayed with 10ppm 2,4-D at 10 days interval. In an open cross pollination from male parent plants grown in alternate rows to flowers from treated female plants gave a fair fruit set with an average of 109 seeds per fruit. Though, the number of seeds per fruit produced by this method is less than obtained by hand emasculation, but it will prove more economical and practical for hybrid seed production in bulk.

Rajasekaran (1970) reported sterility in an inter-varietal hybrid between *Solanum melongena* L. x *S. melongena* var. *bulsarensis* Agrikar. In hybridization programme characters such as disease and pest resistance and drought resistance from wild species are used. Gopimony and Sreenivasan (1971) reported that F1 hybrids obtained from a cross between three cultivars of brinjal with wild *Solanum melongena* var. *incanum* had a high degree of heterosis in branch, flower and fruit numbers and tap roots.

Akiba *et al.* (1972) reported bacterial wilt and anthracnose resistance in Nihon Nassu, a Japanese cultivar, due to one pair of dominant genes. Therefore, this can be a good source for breeding disease resistant fruits.

Mutation breeding studies on UV irradiation of the seeds of parent cultivars Baklazhan 12 and Delikates by Popova and Petrov during 1973, revealed that irradiation speeded up growth, improved vegetative development, earliness and total yield in F1 hybrids.

Viswanathan (1973) while studying the hybrid vigour in brinjal, obtained a best economic combination in a cross between Muktakeshi and Banaras Giant. Similarly, Ludilov (1974), reported that fertility of *Solanum melongena* × *S. integrifolium* F1 hybrids can be increased by maintaining the temperature above 21°C during mitosis, flowering and pollination.

Viswanathan (1975) demonstrated natural hybridization between *S. incanum* and *S. melongena* L. and further reported that application of 0.25 and 0.5 per cent colchicines at one to two true leaf stage to the growing point increases fertility.

S. incanum, a spiny, wild growing species with large flowers and purple fruits, on selfing produced fruits which segregated for size and colour, the largest resembling a round *S. melongena* and smallest being similar to *S. incanum*.

Genetic Engineering for the Improvement of Brinjal (*Solanum melongena* L.)

It is an important vegetable crop of South Asia and Mediterranean countries. A lot of insects and diseases markedly reduce its marketable yield. The genetic improvement by conventional plant breeding is impaired due to lack of resistance to pests and diseases in brinjal germplasm. However, in recent years, genetic engineering of brinjal has resulted in considerable success. Gene encoding insecticidal proteins of *Bacillus thuringiensis* have been introduced into egg plant to develop resistance to Colorado potato beetle and shoot and fruit borers. A fascinating instance of genetic manipulation for parthenocarpy (seedless fruit) has been achieved by altering auxin levels in the developing ovules of transgenic egg plants. This mini-review discusses the introduction of disease tolerance and value addition to the fruits via genetic engineering and future prospects using these techniques. The engineered plant expresses a natural insecticide derived from the *Bacillus thuringiensis* (*Bt*), making it resistant to the fruit and shoot borer (FSB), a highly destructive pest. This tiny larvae account for up to 40 per cent of egg plant losses each year in India, Bangla Desh and Philippine and other areas of South and South East Asia.

Bt. Brinjal is a transgenic brinjal created out of inserting a gene (cry 1Ac) from the soil bacterium *Bacillus thuringiensis* in to brinjal, the insertion of the gene into the cell in young cotyledons has been done through Agrobacterium-mediated vector, along with other genes like promoters, makers etc. When infested by the FSB larvae, the *Bt* protein is activated in the gult leading to lyses and death of the larvae. *Bt* Brinjal is found to be effective against Fruit Shoot Borer (FSB) with 98 per cent insect morality in Bt Brinjals shoots and 100 per cent in fruits compared to less than 30 per cent morality in non-*Bt* counterparts. This is said to give the brinjal plant resistance against Lepidopeteron insects like the brinjal fruit and shoot borer (*Leucinodes orbonalis*) and fruit borer *(Helicover parmigera)*.

At least half a dozen publicly funded research teams and Mahyco are working on this project with a view to increase production of Egg plant, reduce costs, reduce toxic chemical residue, on the fruits and improve the income of a million plus farmers. The companies are Maharashtra Hybrid Seeds Company Ltd. (Mahyco). Besides them technology developed by National Research Centre on Plant Biotechnology with Cry Fa1 has been transferred to Bejo Sheetal, Vibha Seeds, Nath Seeds and Krishidhan Seeds Ltd. Indian Institute of Horticultural Research is also developing the FSB resistant Brinjal by incorporating Cry1Ab gene. Rigorous Scientific tests, including toxicity and Allergenicity evaluation as well as nutritional have confirmed that Bt Brinjal is as safe as its non-*Bt* counterparts.

Chapter 5
Varieties

In brinjal, there is a large variation in plant types *viz.* fruit colour, fruit size and shape. The white, yellow, purple, brown, green, striped and variegated varieties of which some are elongated, pear shaped or round are available in different regions of the country. Some varieties owing to their sizes and colours are preferred in the market. The round or egg shaped varieties are grouped under var. *esculentum*, the long slender types are put under var. *serpentinum* and the early dwarf ones under var. *depressum* (Choudhury,1967). Two main types namely round and long are cultivated all over the country, and following are the important cultivars grown in different regions of India. Extensive work is being done in India to develop attractive, high yielding, disease and insect resistant brinjal cultivars

The most widely cultivated varieties (cultivars) in Europe and North America today are elongated ovoid, 12–25 cm wide and 6–9 cm broad in a dark purple skin. A much wider range of shapes, sizes and colours is grown in India and other Asian countries. Larger varieties weighing up to a kilogram grow in the region between the Ganga and Yamuna rivers, while smaller varieties are found elsewhere. Colours vary from white to yellow or green as well as reddish-purple and dark purple. Some cultivars have a colour gradient, from white at the stem to bright pink to deep purple or even black. Green or purple cultivars in white strippings also exist. Chinese egg plants are commonly shaped like a narrower, slightly pendulous cucumber, and sometimes were called Japanese egg plants in North America.

Oval or elongated oval-shaped and black-skinned cultivars include Harris Special Hibush, Burpee Hybrid, Black Magic, Classic, Dusky and Black beauty. Slim cultivars in purple–black skin include Little Fingers, Ichiben, Pingtung Long and Tycoon, in green skin. Louisiana Long Green and Thai (long) Green in white skin Dourga. Traditional, white skinned, egg shaped cultivars include Casper and Eastern egg. Bicoloured cultivars with colour gradient include Rosa Bianca and Violetta Di Firenze. Bicoloured cultivars in stripping including Listadade ganadia and Udumalapet. In some parts of India, miniature varieties of egg plants called Vengan are popular. There are many named varieties of this annual fruit with new forms being developed.

> ☆ Selection of variety is an important consideration in the cultivation of brinjal as the local preference varies considerably from region-to-region and even district-to-district.

☆ A variety with a particular colour and size of the fruit fetching premium price in the market, may be totally rejected in another area.

☆ Since the ultimate aim of the vegetable grower/ farmer is to get maximum return for his produce, judicious selection of the type with reference to market demands is very important in this crop.

Long Fruited Varieties

Pusa purple cluster, H4, Vijay Hybrid, Azad Kranti, Arka sheel, P.B.R-129-5, Sel–1, ARU-20, ARU-1 and Pusa purple long.

Round Fruited Varieties

S-16, P-8, Pusa purple round, Arka Navneet, T-3.

Long Green Varieties

Arka Shirish, Arka Kusumakar.

Annamali

It is a selection from AC-49A variety, released in 1971 from TNAU, Coimbatore. Plant 65.3 cm tall with a spread of 74.3 cm, characterized by semi erect stem. The fruits are some what club-shaped and slightly curved in the middle, deep purple in winter with a characteristic yellowish mark along the calyx border and a few thorns on the calyx surface and lighter in summer. The cultivar is resistant to aphids, is suitable for ratooning and can yield as high as 300–350 q/ ha. The cultivar is preferred in Cuddalore, Villupuram, Vellore, Thiruvannamalai and Chengalpattu districts and Chennai market.

Arka Keshav

Plant is 80–100cm in height and well branched, developed at IIHR, Bangalore. The fruits are red purple in colour with glossy skin, measuring 18–20 cm in length and 5–6 cm in diameter. The fruits have tender flesh, slow mauring seeds and free from bitter principles. The fruits are borne in two flushes in clusters. Each fruit weighs on an average 50 g. The variety is resistant to bacterial wilt. It yields 300–400 q/ha.

Arka Kusumkar

It is an improvement over a local collection (IIHR-93) from Karnataka and identified in 1981 for cultivation. It produces small, finger shaped pale green good textured fruits with less seeds, that are borne in clusters of 5–7. It is good for cooking. The skin colour is light green. It is a dwarf high yielding variety, bearing 70–75 small fruits in about 110–120 days of transplanting with an average yield of 490 q/ha.

Arka Navneet

F1 hybrid, is a cross between IIHR 22-1 × Supreme (an Australian variety) released by Indian Institute of Horticultural Research, Bangalore and has been identified for cultivation in 1981. Plants are indeterminate. A high yielding hybrid with attractive deep purple fruits and green calyx. The fruit are round to slightly oval, big in size, each weighing 400–500 g. The skin is attractive, with soft and white flesh having few whit seeds. It has excellent cooking qualities. The fruit is free from bitterness and weighs about 398 g. The hybrid yields 650–750 q/ha in 120 days after transplanting.

Arka Nidhi

It is also IIHR, Bangalore product. The plants are tall (90 cm) well branched and compact. Fruits are medium long (length 20c m, girth 9 cm) with blue black glossy skin, green purple calyx, tender flesh with slow maturing seeds, free from bitterness. The fruits on an average weigh 43 g and are borne in clusters in two flushes. Fruit has a good cooking and keeping quality and can be stored up to 20 days at 25 to 30°C. It is bacterial wilt resistant cultivar and yields about 485 q/ha in about 150 days.

Arka Sheel

It is a pure line selection from a coorg type developed at IIHR, Bangalore–India. It is improvement over Pusa Long Cultivar of brinjal. The plant is erect and well branched, bearing medium long, tender fruits with attractive deep purple skin colour. It retains the freshness much beyond the marketable stage. The fruit is much fleshy with less number of seeds. It is a high yielding variety and produces on an average 480 q/ha in about 110–120 days of transplanting.

Arka Shirish

It is an extra long cultivar improved from a local type called Irangeri (IIHR-194) brinjal of Karanataka State. The fruits are tender, extra long, thick with light green skin colour. Flesh soft with half of the fruit length towards stem end. The fruit is fleshy of good texture, attractive with less seeds and of better keeping quality. A high yielding variety, with 25–30 fruits per plant. It yields about of 450–475 quintals per hectare in 175 days of transplanting.

Aru-1

Developed through selection at DARL, Pithoragarh, Uttranchal. Plant is 100–110 cm tall, with purple vein and very acute blade tip angle, bearing single/double light purple and long (17–23 cm) fruits per branch. The fruits are ready for harvest within 110–120 days and may last upto 160 to 172 days of transplanting. The average yield recorded is 290 q/ha. The cultivar has been recommended for J & K, Himachal Pradesh and Uttranchal.

Aru-2C

It has been developed through selection at DARL, Pithoragarh, Uttranchal. A bacterial resistant variety, with cylindrical, long, bright, violet fruits, borne in cluster of 4–6. Under proper agro-climatic conditions can yield 310 q/ha. It has also been recommended for J& K, Himachal Pradesh, Uttranchal, U.P, Bihar, Rajasthan, Gujarat, Haryana, Delhi, Madhya Pradesh and Maharashtra.

ARBH-1 (Shyamal)

It has been developed by Ankur Seeds Pvt. Ltd; Nagpur. Plants are semi-spreading to spreading with vigorous growth. Fruits are purple and oblong. Flesh is milky-white. Shelf life is 4–5 days, hence it has good transport quality. The hybrid is tolerant to sucking-borer insects and diseases. First picking starts 100–110 days after transplanting and continues up to 240–270 days with an average yield of 700–750 q/ha.

Aruna

It is selection from a local cultivar, developed at MPKV–Akola, Maharashtra. Aruna is a prolific bearer, producing small, round-oval deep purple plump fruits with light purple skin. Picking of fruits starts from 95 days from transplanting with an average yield of 412 q/ha during summer and 220 q/ha during summer respectively. It has been recommended for Madhya Pradesh and Maharashtra.

Azad B-1

It is a cross between Banarasi Round × 7103. Plant is 60–70 cm in height, stout, erect and compact, with wide and thick leaf lamina. The average weight of large round and purple fruit is 160–180 cm. Shining purple fruit is resistant to little leaf, phomopsis and fruit rot. The fruit is less susceptible to fruit and shoot borer. Normally it bears 10–12 fruits per plant and the average per hectare is 300–350 quintals.

Azad B-2 (KS-224)

The variety has been developed at Vegetable Research Station, CSAU A & T, Kanpur, from a cross between Pusa Purple Round × Arka Sheel. Plant is about 90 cm high, compact bearing 12-20 round purple fruits, with green shining calyx. The fruits are solitary, with medium thick pericarp and dull white with firm flesh. The average fruit weight is 135g and the first picking may start within 56–60 days of transplanting and may last up to 140–150 days of transplanting. The average yield recorded is 300 q/ha. and has been recommended for cultivation for Punjab, Uttar Pradesh and Bihar.

Azad B-3

It is a long fruited high yielding variety having bright glossy purple fruits, with attractive shape and size, has been recommended for Uttar Pradesh, Bihar, Orissa, Punjab, and Andhra Pradesh. It is a semi erect and evenly spread variety with non-spiny green stem. The fruits are medium thick, long, shining, purple, smooth with shining green calyx (Srivastava *et al.*, 2000), with stigmatic end blunt. It has firm flesh, slow enzymic browning, smooth pericarp besides being less seedy and has better shelf life with prolonged freshness and colour retension. It is suitable for cultivation in the rainy season with normal package of practices *viz.* a dose of 100 kg N, 50 kg each of P_2O_5 and K_2O.

Azad Kranti (PPL x BGL)

The variety was developed at Kanpur, Uttar Pradesh by CSAUAT, during 1983. The plant is erect, with narrow leaf lamina and slightly wavy along the margins. Fruits long are uniform, thick, oblong, 15–20 cm long, dark purple with shining green colour and has less seeds. Calyx green slightly tapering at the distal end. It is an early maturing cultivar and is ready for harvest in 100–105 days of transplanting. On an average it produces 300 q/ha and each fruit weighs about 62 g.

ARC-BJ-105

The plant is upright, about 133 cm in height, stem and leaves with green tinge of purple and covered with fine hairs. The purple dark fruits with greenish flesh are long and slender. The fruits size is 56.7 × 3.4 cm, and each weighs about 130g. The average fruit yield is 2.7–3.4 kg per plant. Flowering and harvesting starts within 21 days and 40 days of transplanting. The variety is highly resistant to bacterial wilt.

Arka Neelakant

It has been developed at IIHR-Bangalore-India. The plants are tall (95.5 cm) and compact. The fruit are short (length 12 cm, girth 8.5 cm) with violet blue glossy skin, green purple calyx, tender flesh having slow maturing seeds and free from bitter flesh having slow maturing seeds and free from bitter principle. Each fruit weighs 40g on an average and are borne in clusters in two flushes. Young leaves are dark green with purple leaf base and veins. Stem is purple green. The cultivar is resistant to bacterial wilt and has very good cooking and keeping qualities. It yields 400 quintals per hectare in crop duration of 150 days.

Azad Hybrid

It has been developed at CSA-Agricultural University, Kalyanpur, Kanpur, Uttar Pradesh by a cross Azad B1 × Kalyanpur Type 3 in which the prolific bearing character has been contributed by the Kalyanpur Type 3 while colour, shape and fruit size by Azad B1. The plants are semi–erect with 60–70 cm in spread, branched, leaf lamina green and entire, early fruiting: fruits bold bright purple, round, calyx green, thin non-fleshy and penduncle green. The fruits retain their purple colour for a longer period and possess better shelf life. The average yield is 450 q/ha. It is less prone to shoot and fruit borer infestation.

APAU Gulabi

This variety was developed at the Department of Horticulture APAU, Hyderabad. The fruits are long and pale purple in colour. It can yield 230–250 q/ha in a crop duration of 160–165 days.

APAU Shyamala

Developed at Andhra Pradesh Agricultural University, Hyderabad. The fruits are very small, round in shape and dark purple in colour. It gives an average yield of 150–160 q/ha in 135–150 days.

APAU Bagyamathi

It is APAU production. The fruits oblong in shape and deep purple in colour. In a crop duration of 150 days it can give 230–300 q/ha.

Barahmasi

A long duration cultivar. The plant is 80–100 cm tall with dark leaves. The fruits are soft, less seeded, 9–10 cm long and 8–9 cm in diameter. The variety has the potential to give 250–300 q/ha.

BH-2

Developed at PAU-Ludhiana from a cross between Punjab Neelam × Punjab Barsati. Plants are tall (79.0 cm), erect and thorn-less with green leaves. The fruits are oblong, shining deep purple and large. The average fruit weight is 237.2 g, with 6.55 mm thick pericarp. The variety may give average yield of 442 q/ha.

Bhagmati

It was released by APAU. It is freely branching, erect type and compact in habit. The emerging leaves are purple in colour, with slightly purple stem. Fairly resistant to fruit borer and little leaf virus. Fruits are borne in clusters of 3-6, deep purple with oblong shape and without spines. It comes to harvest in about 45-50 days after planting. Duration of the crop is 140-160 days. This variety withstands better for drought. Also performs better in upland rainfed cultivation. Yields range from 30-35 tonnes/ha.

Black Beauty

It is an introduction from U.S.A. Plants are 60–75 cm tall, erect and sturdy with dark green colour. The round-oblong fruits are 10–12 cm long with dark purplish black colour. The flowering starts from 80 to 90 days after planting.

CHBR-1

This cultivar has been developed at Horticulture and Agro-forestry Research Programme, Ranchi, Jharkhand. Plants are prostrate (spreading stem), medium tall (89.0cm) with 6–9 branches, violet leaves bearing 15 round fruits which are dark-violet, 9.6 cm thick and weighing about 320 g. The average yield recorded is 756 q/ha and has been recommended for cultivation for Punjab, Bihar and Uttar Pradesh.

Co-1

It is a pure line selection from TNAU, Coimbatore. The plants are erect, compact and bushy with green stem and leaves and greenish purple petiole. The oblong, medium sized fruits are pale green with white background. The soft fruit weighing about 50 to 60g, with less seed, has a good keeping quality. The cultivar is moderately resistant to root rot and reniform nematodes. The average yield recorded is 24 t/ha and lasts till 160 days of transplanting. It has yield potential of 20–25 tonnes/ha. The variety is in great demand in Southern districts of Tamil Nadu, such as Tirunelveli, Ramananthapuram etc., and also Karnatak state in India.

Co-2

It is a pure line selection from a local variety 'Varikkathiri' in the village Negamum and released by TNAU-Coimbatore. The plant is erect and compact, with dark purple slightly oblong fruits having dark purple streaks of different lengths and widths under pale green background without spines on the calyx surface. The variety is very popular in districts of Coimbatore and Periyar. The yield recorded is 350 q/ha and the crop duration is 150 days.

COBH-1

It is developed by TNAU–Coimbatore from a cross between ED-45 × Co-2. Plants are 90.8 cm tall, with an average of 4.4 branches per plant. Each plant yields about 55.4 fruits with an average weight of 64 g per fruit and an average size of 8.94 × 12.43 cm. It gives an average yield of 3.5 kg per plant.

DBRS-44

Plants are 60–70 cm tall, bearing 4–5 fruits in a cluster with an average weight of 125 g per fruit. The average yield recorded is 250–300 q/ha.

DBRS-91

It has been bred at IARI- New Delhi. The plants are semi erect, medium tall (65.0 cm) with 6–8 branches bearing small, round and purple fruits. It is a high yielding variety and average yield recorded is 560 q/ha and has been recommended for cultivation for Madhya Pradesh and Maharashtra.

Dudhia

The variety is famous for its milky white fruits which are best for bhartha preparation and is most suitable for cultivation during winter season.

Green Long

The variety has been developed at RAU, Sabour. The plants are up right, medium tall (73.0 cm) with 8-9 branches bearing about 18 long green fruits with an average weight of 135g per fruit. It gives an average yield of 400 q/ha and has been recommended for cultivation for Punjab, Uttar Pradesh and Bihar.

Gujarat Brinjal-6

It is a cross between Doli-5 × Morvi-4-2, released by Gujarat in 1976. Plant is 63.3 cm tall. The flowering usually occurs within 42 days and produces violet black fruits 10.1 cm long and 9.0 cm diameter.

Gulabi (Sel-4)

Plants are erect, having 5-6 primary branches, green stem with purple shade, and young green leaves with light purple shade on mid rib and the petiole. The medium tall, light purple fruits are borne in clusters of 3-5. Harvesting starts from 60-65 days of transplanting, and the brinjals can with stand long transportation. The variety has been recommended for cultivation states of Chhattisgarh, Orissa and Andhra Pradesh.

Hissar Pragati (H-7)

Variety H-7 has been developed at CCHAU- Hissar through pedigree method from the cross R-34 × Sel-26. Plants are erect, compact, bushy, dwarf and indeterminate. The green foliage turn purple at latter stage. The flowers are whitish purple with green sepals and peduncle. While as fruits are long, dark, bright purple with creamy white flesh picking of fruits starts 45–50 days of planting. The average yield recorded is 334 q/ha and the picking can lasts up to 150 days from transplanting. Variety is tolerant to little leaf and can with stand high temperature. It has been recommended for cultivation in Punjab, Uttar Pradesh, Bihar, Rajasthan, Gujarat, Haryana and Delhi.

Hissar Jamuni

It very popular in Haryana and is a cross between Aushey a round brinjal and R-4 a long brinjal. The medium tall plant is semi spreading, indeterminate, thorn less with purple flowers and foliage. The oblong, 10-15 long fruits are bulged at the blossom end. The fruits are deep bright purple blackish in colour with green purple sepals and the peduncle. The fruit maintains its purple colour till its maturity. Hissar Jamuni is suitable for summer as well as for spring seasons. It is also a good ratoon crop especially for early spring. The variety performed well under normal conditions and records an average yield of 320 q/ha (Kalloo, et al., 1993).

Hissar Shyamal (H-8)

Developed at CCSHAL- Hissar, through back cross-pedgree method from the cross Aushey × BR-112. Plants are spreading with green leaves at the seedling stage. Early, purple green foliage from flowering onwards. Flowers are purple. Fruits are round, dark bright purple with creamy white flesh and with dark green purple pedicel. Picking can start 50–60 days after transplanting and can yield 325 q/ha in 150 days of duration. It has been recommended for cultivation in West Bengal, Assam, Punjab, Uttar Pradesh, Bihar, Chhattisgarh, Orissa, Andhra Pradesh, Rajasthan, Gujarat, Haryana, Delhi, Madhya Pradesh, Maharashtra, Karnataka, Tamil Nadu and Kerala. Tolerant to little leaf and resistant to bacterial wilt.

H-4

It is a cross between Hyderpur × Pusa Purple Long. Released by Haryana Agricultural University, Hissar during

1975. An early variety, the plants of which are semi-erect in habit with spines on leaves, but not on calyx. Fruits are bright dark purple, medium to long, oblong with light green flesh. The first picking starts from 50 days of transplanting.

Jamuni Gola

Variety has been developed at PAU-Ludhiana during 1986. It is a medium tall, spreading, thornless variety. It is a early maturing, bearing round, plump shining purple fruits. First picking starts from 65 days after transplanting. It is selection from local germplasm, S 16.

JB-15

The variety bred at JNKV-Jabalpur. Plants are erect 83.9 cm tall with 7-8 branches having green leaves. It bears 13–15 violet purple fruits per plant which are 14.8 cm long, each weighing about 270 g. Average yield recorded is 300 q/ha, and is recommended for cultivation in Jammu & Kashmir, Himachal Pradesh and Uttaranchal.

JB-64-1-2

It has been bred at JNKV-Jabalpur. The plants are up right, 5–8 branched, bearing 27 small (6cm) round, purple fruits each weighing about 95g. It gives an average yield of 315 q/ha, and has been recommended for cultivation in Madhya Pradesh and Maharashtra.

Junagarh Long

It has been developed from a cross between Doli (female) × in bred Panch Mahal (male) in state of Gujarat in 1967. Plant has semi spreading habit, non-spiny and about 120 cm tall with dark green leaves having wavy margins. The fruits are 17 cm long, 8 cm in diameter, tender, purple and tapering at the distal end. About 20–25 pickings can be taken at short intervals. It is tolerant to major pests *i.e.* shoot and fruit borers and collar rot disease. Picking of marketable fruits starts from 50 days after transplanting. It gives an average yield of 270 quintals per hectare.

Kalpataru (MBH-10)

It has been developed by Maharashtra Hybrid Seeds Co. Ltd. at Jalna. The plants are erect, bushy, compact and spiny bearing round to slightly oval shining, reddish purple fruits with white strips having an average weight of 60-70g per fruit. Calyx is green and spiny. Harvesting starts from 70-75 days of transplanting, and gives an average yield of 400-500 q/ha.

KKM

A pure line selection from Kulathur Local-Tirunelveli by Tamil Nadu Agriculture University Coimbatore, developed mainly for cultivation in irrigated and non–irrigated areas. The plants are compact bearing white small sized, egg shaped fruits borne in clusters of 2–4 with green calyx. It gives an average yield of 370 q/ha in 130–150 days of crop duration. Variety is preferred in Southern parts of the state.

KS-331

This variety has been developed by CSAUA & T-Kanpur and cultivated in Punjab, Uttar Pradesh, Bihar, Chattisgarh, Orissa and Andhra Pradesh. The plants are erect, 70.0 cm tall with 3-5 branches, bearing 16 fruits, each 15–19 cm long weighing on an average 218 g.

Krishnagar Purple Round

It is recommended for West Bengal State. It is a popular variety well suited for growing during winter season and the fruits of which are ready for harvest in about 75–80 days after transplanting.The fruits are large in sized, round, slightly oval, dark purple, fleshy with very less seeds. The average yield varies from 250–350 q/ha.

Krishnagar Green Long

Released by Department of Agriculture, West Bengal. Avery popular variety gets ready in 75–80 days after transplanting. Fruits are 25–30 cm long, fleshy with scanty seeds. It yields 250–350 quintals per hectare. It serves best when fried and is recommended for growing during winter.

Kasi Taru (IVBL-9)

Recommended for U.P., Bihar and Jharkhand. Identified through State Varietal Release Committee.

Kashi Prakash (IIVR Sel-9)

Also recommended for U.P., Bihar and Jharkhand. Identified through Central Varietal Release Committee.

Kashi Sandesh (Hyb)

High yielding hybrid, recommended for all brinjal growing regions of the country. Identified through Central Release Committee.

Kashi Ganesh (Hyb)

Recommended for U.P., Bihar and Jharkhand. Identified through Central Varital Release Committee.

KT

A medium Tall variety, developed at Katrain–IARI Center. Bears 3-5 tender, cylindrical, firm and purple fruits in clusters. A wilt resistant variety, giving an average yield of 371 q/ha.

KT-4

Developed at IARI regional station Katrain of IARI. Its foliage and stem is slightly violet tinged with purple midrib. The cylindrical fruits are borne in clusters of 3–5. Purple fruits are tender with a shelf life of 3–5 days. The cultivar is resistant to bacterial wilt. The average yield is 370 q/ha.

Long-13 (Indo-American Hybrid Seed Co.)

It is an early maturing variety. Plants are up-right, tall with spiny stem. The fruits are dark purple and 15–18 cm long.

MDU-1

It has been developed by TNAU-Coimbatore from Kallampatti Local Cultivar for tropical plains of the state. Plant is vigorous, compact and spreading with broad green leaves with out pigments. Fruits are round, bright purple each weighing 200–250 g. On maturity the purple fruit colour fades to pale pink. It is preferred in some districts of the state *viz.* Madurai and Trichy. The average yield recorded is 340 q/ha in crop duration of 135–145 days.

Manjri

It is a selection from a local material, recommended by Department of Agriculture, Maharashtra State. Fruits are medium sized, round rosy in colour with white strips and very tasty. They are soft but watery in taste. Average yield recorded varies from 275–300 q/ha. It is resistant to bacterial wilt. It is preferred in Madurai and Trichy districts of Tamil Nadu and also in Karanataka state.

MHB-1

Developed by Maharashtra Hybrid Seeds Company Jalna (MAHYCO). This is a tall erect growing hybrid. The fruits are long, blackish purple and cylindrical in shape. Spines are found on the leaves and calyx. Fruits have less quantitry of seeds. It has an yield potential of 500–600 q/ha. This types of fruits are preferred in Karnataka state.

MHB-9

This is another hybrid developed by MAHYCO, Jalna. The fruits are narrow, elongated and very light green in colour. It is also preferred by people of Karnataka state. It has an yield potential of 250–300 q/ha.

MHB-10 (Kalpataru)

It has been developed at Maharashtra Hybrid Seeds Co Ltd; Jalna. Plants are erect, bushy, compact and spiny. Fruits are round to slight oval, shining, greenish purple with white streaks (bicolour) and green spiny calyx. The average fruit weight is 60–70 g. Fruit picking starts 60–65 days after planting and continues up to 125–130 days. Its average yield is 400–500 q/ha.

MHB-20 (Kalpatharu)

It is a product of MAHYCO, the fruits of which are round with bright reddish purple strips on a pale whitish green back ground. Its calyx are bright green with spines. The yield potential is 400–500 q/ha.

MHB-39 (Ravaiyya)

It has been developed at Maharashtra Hybrid Seeds Co. Ltd; Jana. Plants are erect, compact and non-spiny. Fruits are oval, shining, reddish purple and non-spiny. Calyx bears in bunch of 4–6 fruits. It is an early-bearing hybrid and takes about 60–65 days for first picking, which continues upto 125–130 days. It gives an average yield of 350q/ha.

Narendra Hybrid Brinjal-6 (NDBH-6)

It has also been developed by NDUAT (Faizabad)-UP. Plants are non-spiny, semi-erect with green stem having light purple upper portion. Green leaves with rough surface and purple veins. Oblong fruits are purple, soft, each weighing 150-200 g and are borne in clusters. Variety is moderately resistant to shoot & fruit borer, phomopsis blight and little leaf. Average yield recorded is 600 q/ha and has been recommended for the plains Punjab, Bihar and Uttar Pradesh.

Narendra Hybrid Brinjal-1 (NDBH-1)

F1 hybrid, developed by the Narendra Dev University of Agriculture Science & Technology Faizabad–UP from a cross PBR 91-1 × K 202-14 and released by the UP State in 1995. The variety has been recommended for Punjab, Bihar and Uttar Pradesh. The plants are non spiny, semi-erect with green hardy stems having light purple upper portion. The purple green leaves have rough surface with purple veins. Purple oblong fruits are attractive, shining, borne in clusters, are soft textured, less seeded, fleshy with good flavour. Each fruit weighs about 150–200 g. It is a high yielding variety and moderately resistant to shoot & fruit borer, phomoposis and little leaf. Fruits are harvested 70–75 days after transplanting, can give yield of 600 q/ha under normal cultural practices.

Neelam

It is an early variety, having attractive round purple fruits and least resistant to shoot and fruit borer. The cultivar has been developed by PAU-Ludhiana, Punjab. It yields about 300 q/ha.

Pant Ritu Raj

It is a cross between Kalyanpur T-3 × Purple Cluster and recommended for cultivation throughout the country. The cultivar is suitable both for kharif and summer seasons. Plants are 60–80 cm tall, having semi-erect with spreading growth habit. Foliage is green with occasional light purple colour on leaves. The fruits are round, soft and purple with slightly tapering towards the bottom are seedless, good flavoured and have good keeping quality. Picking generally starts from two months of transplanting. It is a bacterial wilt resistant variety, with an average fruit yield of 403 q/ha. Suitable for planting both in autum and summer seasons.

Pant Samrat

It is developed through pure line selection (from local germplasm) at Pant Nagar Agricultural University and has been recommended for cultivation throughout the country. Plants are 80–100 cm tall, robust with broad green leaves. The soft, attractive, purple medium long fruits are borne in clusters. Picking starts from 70 days after transplanting. The variety is resistant to phomophsis blight and bacterial wilt and is less affected by shoot and fruit borer and Jassids. Released by UP State during 1983. It yields about 450 quintals of fruits per hectare.

Pant Brinjal Hybrid 1

Developed by GBPUAT–Pantnagar by crossing PB 129 × PB 225. It is a medium tall variety, having dark green leaves with purple tinge on young leaves. Fruits are long, borne in clusters with soft flesh. The cultivar is resistant to bacterial wilt. Released by UP State in 1993.

PKM-1

The cultivar has been developed by Tamil Nadu Agriculture University, for cultivation in rainfed districts of Madurai, Dindigul, Manner and Thirumalai. It is an induced mutant of a local type called Puzhuthi, the fruits of which are small and slightly oblong obvate with green strips. The fruits produced are 6 cm long and 8 cm in girth. It is a drought resistant cultivar suitable for long transportation. The average yield recorded is 347.5 q/ha in a duration of 150 to 155 days.

PLR-1

It is developed by TNAU-Coimbatore, is a re-selection from a Nagpur ecotype and released during 1990. It is suitable for growing in Coimbatore in all the seasons. It has a longer shelf life *i.e.* up to 10 days under ambient temperature. The fruits are small to medium in size, sometimes borne in clusters, egg shaped, with bright glossy purple fruits. The fruit has high ascorbic acid content of 12 mg/100g fruit weight. It yields on an average of 251 q/ha.

Punjab Bahar

It is a selection, developed from the local material, by PAU-Ludhiana. Plants are 93 cm tall and non-spiny having deep purple foliage with green tinge. Round, dark purple shining fruits, each weighing 200 to 300g are picked 69 days of transplanting. The fruits are less seeded and suitable for bhartha making a special dish prepared in Indian households. Average yield varies from 350-400 q/ha.

Punjab Barsati

Developed by PAU-Ludhiana, a dwarf plant for rainy

season. A prolific bearer having oblong fruits of bluish purple colour. The plants grow to a height of 67–70 cm. The variety so developed is tolerant to shoot and fruit borer but is susceptible to anthracnose. Gives on an average a yield of 350 q/ha.

Punjab Chamkila

Released from PAU-Ludhiana, very suitable for summer and autumn seasons. The fruits are long, thin and dark purple in colour. Variety can yield as high as 400 q/ha.

Punjab Haryana Brinjal-4

It has been developed from a cross between Hyderpur × Pusa Purple Long, as the name indicates, it is the derivative of cross Hyderpur × PPL under the selection No.S-4 done at Punjab Agriculture University, Ludhiana and selection No. H-4 done at Haryana Agriculture University, Hissar. It has been recommended for cultivation in northern states of India. Plants are bushy with pigmented stem bearing long to medium, thin and dark purple fruits with light green flesh.

Punjab Moti

It is a small round, shining dark purple fruited cultivar developed by PAU-Ludhiana for autumn and spring seasons, giving an average yield of 300 q/ha.

Punjab Neelam

It is PAU Ludhiana developed variety, producing oval, medium, shining, dark purple fruits It is suitable for February and August transplanting, giving an average yield of 350q/ha.

Punjab Sadabahar

It is a shoot and fruit borer tolerant cultivar developed at PAU-Ludhiana for rainy season. It has long, thin and deep purple fruits. It gives an yield of 350–400 q/ha and is earlier than Punjab–8.

Punjab–8

It is a highly branched and prolific bearer variety, developed at Punjab Agricultural University–Ludhiana. It is an early and rounded fruited variety bearing medium sized, light purple fruits. The fruiting starts about 65 days after transplanting with an average yield is about of 250–300 q/ha.

Pusa Ankur

It has been developed at Indian Agriculture Research Institute–New Delhi and is recommended for cultivation in Gujarat, Maharashtra, and Central Plains of India. The plants are non-spiny, semi erect, with light purple pigmentation on mid ribs and veins of the younger leaves. Light pigmentation is also visible on younger branches. The fruits are small, oval round, dark purple, glossy and very attractive, each weighing 60-80g. It is an early variety, and first picking can be as early as 45 days of transplanting.The fruits do not loose tenderness and colour even on delayed picking. (Gupta *et al.*, 1999). On an average, Pusa Ankur gives an yield of 30–35 tonnes *per hectare*. Apply two top dressings of urea @ 60–70 kg at one months interval starting from 2–3 weeks after transplanting for better results.

Pusa Anmol

It is a hybrid cultivar with attractive dark purple oblong fruits, derived from a cross between Pusa Purple Long and

Hyderpur at IARI–New Delhi. It is an early cultivar and gives about 80 per cent more yield than Pusa Purple Long.

Pusa Anupam

Developed at Regional Station of the IARI-Katrain through hybridization of Pusa Kranti and Pusa Purple Cluster, during 1991. The plants are medium sized and bushy with foliage and stem colour slightly violet tinged. The cylindrical, medium long, tender, firm and purple fruits are borne in clusters of 3–5, each measuring about 14–18 cm in length, 14–15 cm in girth and 3.4–4.5 cm in diameter. It produces on an average 30–35 fruits per plant. The first picking starts from 135–145 days of sowing. Seeds in the fruit are fewer in number but longer than Pusa Purple Cluster and shorter than Pusa Purple Long. Brinjal is available from August to November in hilly regions and the fruit is suitable for pickle making. It gives an average yield of 300-400 q/ha (Verma *et al.*, 1992).

Pusa Bindu

Plant is pigmented, straight with moderate branching habit and free from spines. The stem is violet-purple with pigmented peduncle. The leaves are 12 cm long and 8 cm in breadth with lobing on tips. Flowers are in clusters with pale-violet corolla. Fruit is pendent, small, oval-round, dark violet purple and glossy. Fruits are solitary to partial cluster and about 14 fruits make one kg. First picking starts from 85–90 days after sowing. An average yield of 300 q/ha was recorded by Kalda and Gupta, during 1990.

Pusa Hybrid–5

It is a long fruited hybrid developed by IARI–New Delhi, the plants of which are vigorous, non-spiny with semi-erect branches. The leaves have bright pigmentation towards the growing tips and the younger leaves. The fruits are medium long, glossy, attractive, dark purple with partially pigmented penduncle, with an average weight of 250–300g. Picking starts from 85–90 days of sowing and continue from October to December. It can yield as high as 460 q/ha (Kalda and Gupta, 1990).

Pusa Hybrid 6

Developed at IARI New Delhi. Plants semi- erect, vigorous, non-spiny,with partial pigmentation on younger leaves and growing tips. The fruits are round, glossy with attractive purple colour with partially pigmented peduncle. The average weight of fruit 250–300 g. It takes 85–90 days from sowing to first picking. It is an early bearing hybrid which has a potential yield of 450 q/ha.

Pusa Hybrid-9

It is a round fruit hybrid with desirable quality characters. The plants are non-spiny, have strong upright branches with light pigmented young leaves. The fruits are big, oval-round, glossy, highly attractive, dark purple with partially pigmented stalk. Each fruit weighs about 300 g and the first picking can start from 85–89 of sowing (Gupta *et al.*, 1999). It is suitable for cultivation in central India, Gujarat and Maharashtra as a main crop. Highest yield recorded is 67 tonnes per hectare where as average yield is 500–600 kq/ha.

Pusa Kranti

It has been developed from a cross between (PPL × Hyderpur) × Wyand Giant. Plants are medium tall, up right, erect in growth habit, young leaves and branches have purple pigmentation. The leaves are dark green and non-spiny. Oblong,glossy, stocky, dark purple fruits are 15–20 cm long, with shining green calyx. Leaves are long, narrow with highly serrated and light green in colour. Stems are medium thick and half green, half purple in colour. The fruits have less seeds and do not touch the ground. On an average each fruit weighs about 68g with a bearing of 22 fruits per plant. The average yield recorded is 400.0 q/ha. The variety is suitable for growing both during spring and summer seasons.

Pusa Purple Cluster

The variety has been developed by IARI, Regional Research Station Katrain after purification from indigenously collected material (Gill *et al.*, 1978). Deep purple medium sized fruits are in clusters of 4–9, which are 10–12 cm long. Plants are tall, medium early, erect and sturdy. leaves and stem have purple pigmentation. Leaves are purple and non-spiny. It is an early maturing cultivar becoming ready for picking in 75 days after transplanting. A heavy yielder and fruiting period can be extended for a longer period if regular pickings are made. PPC is suitable for southern and northern hills of India. The cultivar is moderately resistant to bacterial wilt. Yields about 400–450 q/ha.

Pusa Purple Long

It is an old variety, developed and released IARI- New Delhi. This is a selection from the mixed Baria cultivar grown in Punjab, Delhi and western Uttar Pradesh. The dwarf plants are semi-erect in habit. It is an early maturing cultivar becoming ready for picking in 100 to 110 days. Fruits are 25–30 cm long, purple, glossy and tender. It is suitable for summer and autumn seasons and has performed well in northern India. It has an yield potential of 25–37 tonnes/ha. It is susceptible to bacterial wilt.

Pusa Purple Round

Variety has been developed at IARI-New Delhi. Plants are tall with thick stem of greenish purple colour, with serrated deep green leaves. Fruits are attractive, round with purple colour. It produces 6 fruits per plant with an average weight of 130–140g. It is a shoot and fruit borer and little leaf resistant variety.

Pusa Upkar (DBR-8)

It is a semi vigorous and spineless plant with moderate branching. When mature its stem and leaves are green but show light pigmentation at early stages of growth. Leaf length and width are 14.5 and 8 cm respectively with intermediate lobbing on tip. Flowers appear solitary initially but in clusters later stage with bluish–violet colour. Each medium sized and glossy dark purple fruits weigh about 200g. The average yield recorded is 400 q/ha (Kalda and Gupta, 1990).

Pusa Uttam

Plants are semi-vigorous, up right, well branched and free from spines. Mature plants appear green with occasional light pigmentation on growing shoots. Leaves are 15.5 cm long and 8 cm in width, with intermediate lobing on tip. Flowers with bluish-violet crolla appear in clusters. The fruits are pendent, solitary, oval, medium, large sized, glossy with dark purple skin and green peduncle, each weighing about 250–300 g. The average yield recorded is 410 q/ha (Kalda and Gupta, 1999).

Pusa Bhairav (11a × PPL)-2-4-8-2)

It is a early variety and non-spiny. First picking starts on an average within 85 days of seed sowing for vegetable purposes. Plants is low to medium in height and green leaves are without the spines. Fruits are deep purple, glossy, attractive,12–15 cm long, with pistil end. It is profuse bearer and is resistant to phomopsis fruit rot and blight to an extent of 83 per cent and resistance to disease is governed by recessive poly genes (Kalda *et al.*, 1977).

Ram Nagar Giant

Variety is very popular in Varanasi and its surrounding regions. Plants are 100–150 cm tall, erect, with upright branches having green foliage. Light green soft fruits are 15–20 cm long and 12–15 cm in diameter each weighing 1.5–2.5 kg. The fruits have less seeds, are therefore used for making Bhartha–a favourite vegetable preparation adorning Indian kitchens. The yield recorded is 300q/ha.

Rajindra Baingan

It is a tall variety with strong frame and light green leaves and stems. It is a heavy bearer producing long light green slightly curved fruits which are less seeded. It is moderately resistant to wilting and phomopsis. Is recommended for growing in rainy and winter seasons. Can yield 250-325 q/ha. It is released from Rajendra Krishi Vishwa Vidyalaya, Sagar for South India.

Ravaiyya (MHB-39)

It has been developed by Maharashtra Hybrid Seeds Co. Ltd. Jalna, having erect, compact and non-spiny plants, with oval, shining, reddish purple fruits. The fruits with non-spiny calyx are borne in clusters of 4-6. The early bearing hybrid is ready for first picking in about 60-65 days and may continue up to 125–130 days. The average yield recorded is 350 q/ha.

RHRBH-1

It is a cross between RHRB-1 and JB-16, developed and released by MPKV-Rahuri. The purple, non-spiny fruits are 8.5 cm in length and 6.4 cm in diameter, with green calyx and purple white strips. It gives an average yield of 576 q/ha.

RHRBH-2

It is a cross between RHRB-1 and RB-25, developed at MPKV-Rahuri. Plants are short and spreading producing purple fruits with white strips. The fruits are 5.5 cm long and 4.5 cm in diameter with spiny calyx. Can yield as high as 407 q/ha.

RHRBH-3

It is also a cross between RHRB-1 and Vaishali, developed at MPKV–Rahuri, has recorded an yield of approximately 440q/ha. The variety so developed has tall and spreading plants, with purple fruits having white strips. The fruits are 7.5 cm in length and 5.5 cm in diameter with spiny calyx.

Round-14 (Indo-American Hybrid Seed Co.)

Plants are tall, spineless. Fruits are medium sized, round, light purple in colour with prominent scar at the blossom end.

Shyamal (ARBH-201)

This spreading and vigorous hybrid has been developed at Nagpur by Ankur Seeds Pvt. Ltd. The oblong and purple fruits of the hybrid have milky white flesh. Its fruits has a good keeping quality even at room temperature and stand long transportation. Fruiting starts from 100-110 days after transplanting and continues up to 240–270 days, and the average yield recorded is 700–750 q/ha. It is resistant to sucking borer, and hybrid been recommended for Punjab, Haryana, Rajasthan, Gujarat, Maharashtra, Bihar, plains of Uttar Pradesh, Andhra Pradesh, Orissa, and eastern parts of Madhya Pradesh for cultivation.

Surya (SM 6-7)

It is a non-prickly, spreading and bacterial wilt resistant variety developed at college of Horticulture, Vellanikkara and recommended for cultivation in Karnataka and Tamil Nadu, in addition to Kerala. The purple fruits are oval in shape. Average yield recorded is 200 q/ha in a span of 150 to 180 days.

Swetha (SM 6-6)

It is a bushy and bacterial resistant cultivar developed at college of Horticulture, Vellanikkara- Kerala. It bears purple flowers with tinge on leaf stalk and medium long, solitary, light green fruits occasionally in clusters. It can give an average yield of 300 q/ha in about 135 to 150 days. It can be grown in seasons of May–August, September–December and January and has been recommended for cultivation in Jammu & Kashmir, Himachal Pradesh, Uttaranchal, Rajasthan, Gujarat, Haryana, Delhi, Karnataka, Tamil Nadu and Kerala.

Swarna Shree (CHES-157)

It has been developed through hybridization followed by pedigree selection and back cross method at Horticulture and Agro-Forestry Research Programme, Ranchi-Jharkhand.

The round-oval, creamy white and soft fruits are preferred for making Bhartha a special curry very common in Indian families. The flesh of fruits is spongy with less seeds. It is a high yielding and moderately bacterial wilt resistant variety. It can yield as high as 600-650 q/ha in a duration of 130-150 days. It is popular in Bihar, Jharkhand and its adjoining areas.

Swarna Mani

It has been developed through hybridization followed by pedigree selection and back cross method at Horticulture and Agro-Forestry Research Programme, Ranchi-Jharkhand. The round purple shining fruits are attractive and moderately resistant to bacterial wilt. A record fruit yield of 600–650 q/ha has been obtained.

Swarna Pratibha

It is a pure line selection with medium long purple shining fruits. Each fruit weighs about 200g. It is a bacterial resistant cultivar recommended for cultivation in Bihar, Jharkhand and its adjoining areas. Average yield recorded in 130–150 days duration is 450–500 q/ha.

Swarna Shyamli

It has been developed through pure line selection at Horticulture and Agro-Forestry Research Programme, Ranchi, Jharkhand. The creamy white, round-oval and soft fruits are preferred for preparation of Bhartha. It is bacterial wilt resistant cultivar, recommended for cultivation in Bihar, Jharkhand and its adjoining areas. The yield potential of this cultivar is 550–600 q/ha.

S-1

It has been developed by Punjab Agricultural University, Ludhiana. Fruits are round, deep shining purple in colour with fruit weight of 200–250 g. The average yield recorded is 180–200 q/ha.

Swarna Shakti

Resistant to phomopsis blight, moderately resistant to bacterial wilt. Recommended for Bihar and Jharkhand. Identified through State Variety Release committee.

Swarna Ajay

Resistant to bacterial and phomopsis blight. Recommended for Punjab, Uttranchal, Bihar, Jharkhand and Uttar Pradesh. This cultivar is also identified through State Variety Release Committee.

Swarna Shobha

Tolerant to bacterial wilt. Recommended for Bihar and Jharkhand. Identified through State Variety Release Committee.

Surati Gota

It is a selection from local variety, released by Department of Agriculture, Maharashtra State. Fruits are medium big, round with purple colour. Soft and watery in taste. It yields 180–200 quintals of fruits per hectare.

Suphal

This is an early maturing F1 hybrid. The fruits are of good quality, deep purple, oblong, oval glossy about 500g in weight. This variety can be grown throughout the year in South India. The average yield recorded is around 600 quintals.

Type-3

It has been developed at Vegetable Research Centre, Kalyanpur, Kanpur- UP. The plants are spreading type. The fruits are round, light–purple with light green flesh and of good quality. Each weighing about 300-400g. It can be grown in Spring–Summer season also. Average yield recorded varies from 250–300 q/ha.

Utkal Jyoti (BB-13)

It has been developed at Orissa University of Agriculture and Technology Bhubhaneshwar, derived through pedigree method from a cross between KT-4 × BB-11. Plant is a non-spin and tall. Fruiting is profuse but in clusters. Fruits are long, purple, small medium with white flesh and have long shelf life. It is bacterial wilt resistant variety, recommended for cultivation in Karnataka, Tamil Nadu and Kerala. The average yield recorded is 348 q/ha in span of 140-180 days.

Utkal Keshari (BB-26)

It has been developed at Orissa University of Agriculture and Technology Bhubhaneshwar, derived through pedigree method from a cross between BB-11 × KJ-3-1. Plant is a tall erect with pigmentation on branches and leaves. Fruits are long, medium thick, basal portion slightly broad, deep purple with 5–6 days of shelf life. It can yield 322 q/ha in about 130–140 days of transplanting. It is a bacterial and fusarium wilt resistant cultivar, and is recommended for cultivation in Gujarat, Rajasthan, Haryana, Delhi, Madhya Pradesh and Maharashtra.

Utkal Madhuri (BB-44)

It is a bacterial resistant cultivar recommended for

cultivation in Chhatisgarh, Orissa, Maharashtra, Andhra and Madhya Pradesh. It has been developed Orissa University Agriculture and Technology-Bhubaneshwar, through pedigree method from the cross PBR125-5 × Pipli–4 (local collection). Plants are medium tall with green stems and leaves, bearing solitary fruits that are long, green (basal portion white) with white flesh. Fruits have longer shelf life of 7- 8 days and are ready for first picking within 100-105 days after transplanting. It is resistant to fusarium wilt, and drought and has long fruiting period. It gives an average yield of 316 q/ha.

Utakal Tarini (BB-77)

It is a bacterial resistant cultivar recommended for cultivation in Chhatisgarh, Orissa, West Bengal, Assam, Maharashtra, Andhra and Madhya Pradesh. It has been developed by Orissa University of Agriculture and Technology-Bhubaneshwar, through pedigree method from the cross Pusa Kranti × Gopal Local. Plants are medium tall with purple pigmentation on leaves and stem, bearing solitary fruits (occasionally two fruits/cluster), that are oblong, medium sized, purple with creamy white flesh. Harvesting can start from 90–95 days of transplanting. Fruits so harvested have 6–7 days of shelf life and the cultivar can yield about 340 q/ha in 130–140 days of crop duration.

Vijay Hybrid

It is developed from Indo-American Hybrid Seeds Ltd. It is a high yielding hybrid having purple, oblong to oval fruits with green calyx Fruits weighing approximately 170 g are ready for first picking within 102 days of transplanting. Average yield recorded is 400 q/ha.

VRBHR-1

It is a product of IIVR- Varanasi, and has been developed from a cross between Pant Ritu raj and BR-SPS-14. Plant are 70.8 cm tall bearing round, light purple fruits that are 11.6 cm in length and 10.28 cm in diameter with an average weight of 569 g. The average fruit yield recorded is 789q/ha.

A large number of cultivars and improved varieties differing in shape, size and colour of fruits are grown in India. Since consumer preference varies from region to region and from district to district, therefore, judicious selection of varieties play an important role in successful brinjal cultivation. There are large number of locality specific cultivars like Banaras Giant, Wayanad Giant, Mukthakeshi, Manjiri Gota and Kashmir local etc. some local cultivars exhibit bitterness due to presence of glycolalkaloids like solanin. Solanin content varies from 0.37 to 4.83 mg/100g in most cultivars and concentration of 20mg/100g fresh weight produces bitter taste and off flavour. Varieties also vary for polyphenol oxidase content which imparts brown discolouration when the brinjals are cut open.

Upright varieties with sturdy growth habit, high yielding, fruits with soft flesh, low seeds, low solanin, disease/pest resistant and attractive glossy skin are generally preferred by the consumers.

The other important varieties, with important features like maturity, fruit characters, average yield and region for which they have been recommended are listed below.

Important features of some important varieties are listed in Table 11.

Table 11: Important Features of some Varieties

Variety	Bearing Maturity	Fruit Shape	Colour	Average Yield (tonnes/ha)	Recommended for
ABH 1	Early	Oval	Purple	37	Northern & central India, with parts of Maharashtra & Gujarat
ABH-2	Early	Oval	Purple	39	Do
Annamalai	Medium	Long	Light purple	25	Tamil Nadu
Arka Keshav	Medium	Long	Fruits long red Purple & glossy	45	Mild climate of southern India Resistant to bacterial wilt
Arka Kusumkar	Medium	Long cluster	Small green fruits in clusters of 5-7	45	Mild climate of southern India
Arka Navneet	Early	Oval	Dark purple	60	do
Arka Neelkantha	Early	Long	Short purple fruits borne in clusters of two	40	do
Arka Nidhi (BWR 12)	Early	Long	Resistant to bacterial Wilt, medium long blue black glossy fruits	35	Bacterial wilt resistant in Southern mild climate
Arka Sheel	Medium	Long	Medium long deep shining purple fruits	45	Mild climate of southern India
Arka Shirish	Medium	Long	Extra long fruits with green colour	47	do

Contd...

Variety	Bearing Maturity	Fruit Shape	Colour	Average Yield (tonnes/ha)	Recommended for
Azad B-1	Medium	Round	Purple	30	Uttar Pradesh
Azad Kranti	Early	Long	Purple	25	North & North Central plains
ARU-2C	Early	Long cluster	Violet purple		Hilly regions, Ganga, Sutlej, Ganga Alluvial plains & Karnataka
Bhagymathi	Early	Oblong	Light purple	15	Andhra Pradesh
BR 112	Medium	Round	Purple	35	Haryana
Co 1	Medium	Long	Light	25	Tamil Nadu
Gujarat-6	Medium	Long	Violet black	25	Gujarat
Hissar Jamuni	Medium	Oblong	Dark purple	35	Haryana
Hissar Shyamal	Medium	Round	Dark purple	35	Haryana
Jamuni Gola	Medium	Round	Purple	25	Punjab
Junagarh Long	Medium	Long	Pinkish	25	Gujarat
K 2029	Medium	Round	Purple	35	Uttar Pradesh
Manjri Gota	Late	Round	Bicolour	15	Maharashtra
MDU 1	Medium	Oval	Light purple	25	Tamil Nadu
Mysore Green	Early	Long cluster	Green	25	Karnataka

Contd...

Variety	Bearing Maturity	Fruit Shape	Colour	Average Yield (tonnes/ha)	Recommended for
NDB-25	Early	Long	Dark purple	30	Uttar Pradesh
NDBH-1	Early	Long	Dark purple	52	Northern & central India
NDBH-6	Medium	Round	Dark purple	53	Maharashtra & Gujarat
Pant Rituraj	Medium	Oval round	Violet purple	35	All over India
Pant Samarat	Early	Long Partially cluster	Purple	35	do
PH-4	Medium	Long	Purple	30	do
Punjab Bahar	Late	Round	Purple	20	Spring-Summer crop of Punjab
Punjab Barsati	Medium	Long cluster	Purple	25	Punjab
Punjab Chamkila	Early	Long	Dark purple	25	do
Punjab Neelam	Medium	Round	Purple	30	do
Punjab Sadabahar	Early	Long	Black purple	30	do
Pusa Anupam	Medium	Long cluster	Purple	30	Highly regions & mild climate of southern India
Pusa Bhairav	Early	Long	Dark purple	30	Phomopsis resistant

Contd...

Variety	Bearing Maturity	Fruit Shape	Colour	Average Yield (tonnes/ha)	Recommended for
Pusa Bindu	Early	Oval round	Violet purple	30	Northern planis, Gujarat, Maharashtra
Pusa Hybrid-5	Early	Long	Dark purple	50	North, central & southern states
Pusa Hybrid-6	Early	Round	Dark purple	45	Northern & central India
Pusa Hybrid-9	Early	Round	Dark purple	56	Maharashtra & Gujarat
Pusa Kranti	Medium	Long	Dark purple	35	All over India
Pusa P C	Medium	Long, cluster	Violetpurple	30	Hilly regions & mild climates of southern India
Pusa P L	Extra early	Long	Purple	35	All over India
Pusa Upkar	Early	Round	Dark purple	40	Northern plains, Gujarat & Maharashtra
Pusa Uttam	Early	Oval	Dark purple	40	Northern plains, Gujarat & Maharashtra
Type 3	Medium	Round	Light purple	25	Uttar Pradesh
Vaishali	Medium	Oval	Bicolour	25	Maharashtra
White Cluster	Early	Oval	White	20	Tamil Nadu

Table 12: Important Features of some Varieties Developed by Institutions and Universities

Institutions	Variety	Special Features
IIHR Bangalore	Arka Kusumkar*	Small green fruits borne in clusters of 5-7
	Arka sheel*	Medium long deep shining purple fruits
	Arka Nidhi*(BWR12)	Medium long blue black glossy fruits(B W. Resistant)
	Arka Shirish	Extra long fruits with green colour
	Arka Neelkanth	Short purple fruits borne in clusters of two (B Wilt Resistant)
	Arka Keshav	Fruits long red purple*glossy. Resistant to bacterial wilt
IARI, New Delhi	Pusa Kranti*	Oblong, 15–20 cm long dark purple fruits.
	PPC*	10–12 cm long purple fruits borne in clusters, tolerant bacterial wilt
	PPL*	Long purple glossy fruits
	Pusa Anupam(KT4*)	Cylindrical purple fruits borne in clusters
	DBR*	Round dark purple fruits of 295g
	Pusa Purple Round	Fruits round and purple
	Pusa Bairav	Fruits long & purple. Resistant to Phomopsis
	Pusa Uttam	Early variety with oval dark purple fruits
	Pusa Utkar	Early variety with round dark purple fruits
	Pusa Bindu	Early, Small oval-round violet-purple fruits
	Pusa Ankur	Fruits oval round, dark purple & glossy

Contd...

Institutions	Variety	Special Features
Kerala Agri. Uni.Trissur	Surya (SM 6-7)*	Small purple oval fruits. Resistant to bacterial wilt
	Swetha (SW 6-6)*	Small white elongated fruits. Resistant to bacterial wilt
	Haritha	Long light green elongated fruits. Resistant to bacterial wilt
Tamil Nadu Agri.	Co-1	Oblong, pale green fruits
Uni Coimbatore	Co-2	Oblong fruits having dark purple streaks under pale back ground
	PKM-1	Small ovate fruits with green strips, developed through mutation breeding
	PLR –1	Small to medium sized egg shaped fruits with purple colour
	MDU-1	Large, round & purple fruits
	KKM-1	Small, white, egg shaped fruits borne in clusters of 2-4
Annamalai Uni. Tamil Nadu	Annamalai	Aphid resistant. Fruit oblong, purple, few thorns on the calyx.
GBPUA & T, Pantnagar	Pant Rituraj*	Long purple round fruits
	Pant Samrat*	Fruits long & purple, resistant to bacterial wilt & phomopsis blight, less attack of shoot & fruit borer & jassids
Haryana Agri.	Hissar Pragati(H7)*	Fruit dark purple, tolerant to little leaf

Contd...

Institutions	Variety	Special Features
University, Hissar	Hissar Shyamal(H8)*	Round dark & bright purple fruits
	Hissar Jamuni	Oblong dark purple fruits
Punjab Agri.	Jamuni Gol(S 16)*	Long plumpy & shining purple fruits.
University, Ludhiana	Punjab Barsati*	Medium long & shining fruits, Resistant to fruit borer
	Punjab Neelam	Long purple fruits
	Punjab Sadabahr	Long black purple fruits
	PH-4	Fruits medium to long, thin & dark purple
CSAUA & T, Kanpur	T-3*	Round light purple fruits with white stylar end
	KS-331*	Long purple fruits of 218 g weight
	Azad Kranti (KS701)*	Medium thick & long purple fruits tapering to distal end
	Azad B2 (KS224)*	Solitary round purple fruits of 135g
MPKV, Akola	Aruna*	Fruits round to oval with light purple skin
DARL, Pithoragarh	ARU-1*	Long light purple fruits borne single or double
	ARU-2C*	Cylindrical & violet fruits borne in clusters of 4-6(BWR)
CHES, Ranchi	CHBR-1*	Round dark violet fruits
JNKV, Jabalpur	JB-15*	Long violet purple fruits of 270g weight
	JB-64-1-2*	Small round purple fruits of 95 g weight

Contd...

Institutions	Variety	Special Features
OUA & T, Bhubaneswar	Utkal Tarini (BB 77)*	Resistant to bacterial wilt, medium sized, oblong deep purple fruits
	Utkal Madhuri (BB-44)*	Medium long green fruits, white striped distal end (B W resistant)
	Utkal Jyothi (BB-13)*	Small to medium long purple fruits, bacterial wilt resistant
	Utkal Kesari (BB-26)*	Deep purple fruits, medium large, cylindrical, thick basal portion
RAU, Sabour	Green Long*	Long green fruits of 135g
APAU, Hyderabad	Gulabi(Sel-4)*	Light purple, medium long fruits borne in clusters of 3-5, suitable for long distance transport, very small purple fruits
	Shyamal Bhagyamathi	Oblong & deep purple fruits
PRVV, Akola	Aruna	Small round purple fruits
MPKV, Rahuri,	Vaishali	Fruits oblong, purple with white strips
Maharashtra	Pragati	Oval, purple fruits with white strips and spines on peduncle

* Varieties released/identified by AICRP (Vegetables)

BWR= Bacterial wilt resistant

☆ *Uttar Pradesh*: Black beauty, Banars Giant, T1, T2, T3, and T4.

☆ *Bihar*: Muktakeshi, ST1 and ST2

☆ *Punjab*: Black beauty, P-38 and P34.

☆ *Maharashtra*: Surti Gota, Manjri Gota and American Purple.

☆ *West Bengal*: Krishnanagar purple, Krishnanagar green long.

☆ *Tamil Nadu*: Co-1, Co-2, Annamalai, MDU-1, PKM-1 and Gudiyatham

☆ *J&K*: Kashmir local

☆ *Indo-American Hybrid Seed Co.*, *Bangalore*: Sophal, Long-13, Round-14

Brinjal Varieties Grouped on the Basis of Shape and Colour of Fruits

(i) *Long Purple*: PPL, PPC, Pusa Kranti, Pusa Anmol, Pusa Bhairav, Azad Kranti, Arka Sheel,Junagadh Long, Pant Samrat.

(ii) *Long Green*: Arka Kusumakar, Arka, Shirish, Krishnanagar Green Long.

(iii) *Round Purple*: PPR, Jamuni Gola (Kalyanpur), Suphal, Arka Navneet, Krishnanagar Purple Round, Pant Rituraj, Surti Gota, T-3.

(iv) *Round Green*: Banarsi Giant (UP)

(v) *Round White*: Swarna Shree

(vi) *Round with strips* (purple white): Manjari, Manjari Gota

1. *Oval or oblong* (purple): Arka Navneet, Junagadh oblong, H-4, Pusa Anmol, Suphal (F1 Hybrid).

Brinjal Varieties Grouped on the Basis of their some Morphological Distinctions

1. *Fruit borne in clusters:* Arka Kusumkar, Arka Neelkanth, PPC, Pant Samrat, ARU-2C (almora)
2. *Spines Present:* H-4, Majari, Majari Gota
3. *Plant growth erect:* Pant Samrat, H-4
4. *Plant growth spreading:* T-3, Pant Rituraj, Surati Gota

Brinjal Hybrid F1

A. Long Type

1. Pusa Hybrid-5 (IARI)
2. Pant Hybrid-1 (Pantnagar)
3. ARBH-201 (ANKUR SEED)
4. MHB-1 (MAHYCO)
5. Arka Kusumkar
6. Long-13 (IAHS)

B. Round Type

1. Pusa Hybrid-6 (IARI)
2. Pusa Hybrid-9 (IARI)
3. Vijay Hybrid (CSAU, Kanpur)
4. Azad Hybrid (CSAU, Kanpur)
5. NDB Hybrid (Faizabad)
6. MHB-2 (MAHYCO)
7. MHB-10 (MAHYCO)
8. Round -14 (IAHS)
9. Arka Navneet (IIHR)

C. Oval to Oblong

    1. Suphal (IAHS)

D. Small Sized Fruits

    1. ABH-1

    2. ABH-2

    3. Phule Hybrid-2

Chapter 6
Cultivation

Climate

It is a long warm season crop, adopted to a wide range of climatic conditions from east to west and north to south of the country. It can grow satisfactory up to an elevation of 1200 meters and most cultivars appear to be day length neutral. Brinjal requires warm season during its growth and fruit maturation and is susceptible to severe frost. However, late round fruited varieties are more tolerant to frost than the early long fruited varieties. The cool season affects adversely plant growth, fruit quality in respect to size and colour development. The optimum temperature of 22 to 30°C is most favourable for its successful production of flowers and fruits and its growth is likely to stop at temperatures below 17°C, as low temperature during the cool season causes abnormal development of the ovary (splitting) in flower buds which then differentiate and develop into deformed fruit during that season (Nothmann and Koller, 1973). A diurnal variation in temperature is not generally essential and may limit development although some cultivars respond to night temperatures of 20-25°C. Whereas, seeds germinate well at 22-25°C. A short lived perennial plant, it is usually cultivated as an annual. Being traditionally a summer vegetable, although now it is possible to buy brinjal almost all year because many farmers use green houses. The most satisfactory environmental conditions are normally found in lowland costal areas with relatively little temperature variation. High soil temperatures are often injurious to the root system therefore, cultural practices like mulching are often carried out to deduce variation in soil temperature. The minimum soil temperature for germination of vegetable seeds ranges from 2 to 15°C, however, optimum soil temperature for seed germination in most of the vegetable crops lies between 20°C to 30°C. While maximum temperature for germination of warm season vegetables is 35°C to 40°C. The soil temperature requirements for seed germination in brinjal as reported by Maynard *et al.* (1980) is:

Crop	Soil Temperature			
	Minimum	Optimum Range	Optimum	Maximum
Brinjal	15.5	24.0	32.0	28.0	35.0

Of all the solanaceous vegetables brinjal is the most sensitive to low temperature. In the plains of northern India, where the winter is severe during December–February, its growth is adversely affected due to frost and low night temperature. In hilly regions of the country, it is grown only during summer. It has been observed that round/

hybrids are more tolerant to frost than the early long/ hybrids. Similarly, late cultivars are tolerant to mild frost than the early ones. Nothmann *et al.* (1979), reported during warm season, the growth and flowering are normal and almost all fruits develop from the basal flowers. Whereas in cool season, the growth is slow and the fruit quality is poor both in size and colour. They further observed that colour development of the fruit seems to be related to vigour of the plant growth. Lombardi and Restaino (1981) reported that anthesis of flowers in different cultivars are influenced by both temperature and the genotype. They further observed that the genotype and temperature interact in their effect on leaf size and also effect the flower set.

Season

In hills and Kashmir valley brinjal is normally sown during April and transplanting is done during May, however, early sowing can also be done under protection and only one crop is raised. Where as in plains there are three seasons for growing brinjals *viz.* February–March, May–June and October–November.

☆ Autumn-winter crop: Crop is sown in June and transplanted in July.

☆ Spring–Summer crop: Crop is sown in early November and transplanted in January–February. Due to low temperature, seedlings usually take 6-8 weeks for attaining normal size and the nursery beds are to be protected from frost. The young seedlings are to be protected from winter injury during nights of December and January.

☆ Rainy season crop: Seeds are sown in March–April and transplanted during April–May. Being a low priced vegetable, rainy season crop is the most economical in many parts. In East and South India, brinjal can be grown all the year round, because of the agro-climatic conditions there. However, the main season being July–August.

Soil

Brinjal can be grown practically on all types of soils from light sandy to heavy clay. Soil for growing brinjal should be deep, fertile and well drained, but soils with good moisture retaining properties are generally considered suitable. Sandy or sandy loam soils are good for early crop. Higher yields are obtained when brinjal is raised in clay loam and silt–loam soils. The root system is sensitive to excess soil water and deep cultivation prior to planting is preferable since the plants are normally grown for more than one year. The brinjal plant is moderately tolerant to acidic soil but for better growth and development, the soil pH should be not more than 5.5 to 6.0, where as salinity affects plant growth and yield. Light soils are most suitable for early crop as these get warmed up quickly, but such soils are often low in nutrients and poor in moisture retention, therefore, need heavier fertilizer application and frequent irrigations. Whereas heavier soils are preferred for higher yields but the crops mature late. Such soils have considerable reserve of plant nutrients and also retain moisture for longer periods. Therefore, only smaller quantities of fertilizers are required and frequent irrigations are not necessary. Dig the soil well or plough and harrow the land at least 4-5 times 2 to 3 weeks before planting for better results. Bulky organic manures should be incorporated into the soil during land preparation. When land is sufficiently prepared and leveled, the whole plot is divided into the beds of suitable dimensions before transplanting. Brinjal plant is adopted to both wet and dry season cultivation but excessive rainfall will check both vegetative growth and flower formation.

Nursery Raising

Brinjal is mainly propagated through seed. The seeds are sown in the nursery and seedlings are transplanted in the field. Seeds of brinjal are sown in finely prepared raised nursery beds, which are 20–25 cm high alternating with water channels. A sufficient amount of fine and well decomposed farm yard manure or compost is well mixed into the top soil of the nursery beds a few days before seed sowing. The convenient bed size for raising nursery is 1m × 3m, however the length may vary depending upon the requirement. Organic manures is added @ 10kg/m² at final preparation. Two days before seed sowing, the nursery beds should be thoroughly drenched with captan suspension to avoid occurrence of damping off of seedlings followed by seed treatment with 30 ppm IAA before sowing. About 300–400 g of seed is sufficient to cover one hectare land. Most preferably the seed should be sown thinly in lines 5 cm apart at the depth of 1–1.5 cms. Generally, about 250 seeds make one gram. After sowing of the seed, nursery beds are covered with paddy straw or dry grasses till the seeds germinate. Depending upon the weather conditions water is sprinkled on the straw covered nursery beds. Flooding should be avoided at all times since brinjal seedlings grow fast, sufficient care must be taken to sow seeds as thin or loose as possible, most preferably in rows. Protect the nursery from hot sun, heavy rains and frost. A light dose of Ammonium sulphate or Calcium ammonium nitrate (CAN) may be applied in the nursery by putting one handful of fertilizer in one bucket of water to promote the growth of seedlings at the time of watering. The seeds normally germinate in 12 to 18 days and the germination is epigeous.

Seedlings at 8–10 cm in height with two to three true leaves are ready for transplanting within 4 weeks during summer and 6 weeks during winter. The seedlings should be hardened before lifting for transplanting, as they withstand transplanting shock better and establish well when transplanted in the main field. The healthy and stocky seedlings which free from diseases and shoot and fruit borer infestation should be used for transplanting. Chanule and Pandey (1958) while studying the effect of hardening of seedlings of the variety Manjri Gota, found that the plants hardened at medium and high moistures gave a significantly higher yield than the unhardened plants by 53 per cent to 66 per cent. Similar results were reported by Desai (1962), who found that maximum hardening of seedlings by irrigating after every twelfth day, increased fruit yields. Response of hardening of seedlings varies from variety to variety, as variety Surti Gota and Manjri Gota responded favourably to the treatment, while as American Purple showed no effect on yield. Hardened seedlings withstand transplanting shock better and establish well in main field. The nursery area requirement in transplanting vegetables is calculated based on the following formula:

$$\text{Area for raising seedling} = \frac{N}{S} \times d$$

where, N = Number of seedlings for unit area in the main field.

S: Expected number of seedlings in each row.

d: Distance between two adjacent rows.

Seed Rate

In order to raise brinjal crop on one hectare area, about 300-400g seed is sufficient to provide sufficient seedlings. Brinjal seeds have 70–80 per cent germination.

Field Preparation

Well prepared field helps promote the growth and development of an excessive root system like that of brinjal. The field for growing brinjal should be well drained, fertile, preferably sandy loam. While as poorly drained soils usually result in reduced functional root area, poor plant growth and low yields. As nutrients and water are absorbed through tiny root hairs that extend from the root system therefore, soils should usually be prepared to a fine tilth by 3-4 ploughings, to a depth of 20 cm. At last ploughing 25–30 tonnes of well rotten organic manure should be incorporated in to the soil and field leveled properly and made into beds of convenient sizes.

Method of sowing/Spacing

Brinjal can be direct seeded but mostly the crop is established with transplants. Strong, healthy, robust and hardened, 20–25 days old transplants 15–20 cm tall are grown 60–75 cm apart in single rows spaced 1–1½ m apart, either on flat ground or on raised beds of convenient size for irrigation. The seedlings should be hardened by withholding watering for 4 to 6 days before transplanting. However, nursery bed should be softened with water, if soil becomes too hard, sothat roots are not injured when the seedlings are uprooted. Crumbling of roots should be avoided while transplanting. In high rain fall regions or during rainy season brinjal plants are grown on raised flat beds for proper drainage. In undulating land, in order to avoid soil erosion, small pits are dug at the point of planting and seedlings are planted. The spring–summer crop may be raised on ridges and furrow system for efficient use of water. Spacing depends on variety, season and fertility of the soil. For long duration and spreading varieties, a spacing of 75–90 cm × 60–75 cm and for bushy and non-spreading varieties a distance of 45–60 cm on either side are given. While for early and less spreading varieties, paired row planting is advantageous due to easiness in harvesting and other cultural operations. The long fruited varieties are transplanted at spacing of 60 × 60 cm and round fruited at 75 × 75 cm.

Seth and Dhauder (1970) reported that yield per plant and fruit quality were slightly better with an inter-row spacing of 100 cm compared with 75 cm in cv. Pusa Purple Long. In a three year study by Sencan (1970) with the variety Yesilkoy 27 the best yields were obtained at 60 × 40 cm spacing. According to Ruiter (1974) fruit number was highest with plants spaced at 100 × 50 cm with five branches per plant. But the average fruit size was greater with 100 × 75 cm and three branches per plant. Trials conducted at Kalyanpur showed that the yield from plants spaced at 75 × 50 cm and 75 × 70 cm were significantly higher than 75 × 90 cm.

Water Management

Timely Irrigation is essential especially for fruit set and development. Irrigation requires a relatively stable amount of available soil moisture during the first 70 days for a study growth pattern that is essential for good brinjal production. It can withstand drought later to some extent and still revive upon soil moisture. Irrigation is essential for brinjal cultivation especially in regions, where there is little or no rains during growing season. High yields are obtained under optimum moisture conditions. Brinjal being a shallow rooted crop needs irrigation at frequent intervals, therefore, it is essential that timely irrigations are given for good fruit set, its development and higher yields. Usually during summer after every 4th or 5th day the brinjal crop is irrigated and 10 to 12 days during winter season. Irrigation is given subject to local considerations, such as soil type, stage of the growth and weather conditions. During frost the crop should be given frequent irrigations in order to keep the soil moist. Care must be taken to drain off the excess irrigation or rain water, because it adversely affects the plant growth whereas the moisture stress increases the uptake of N and K. Under hot, dry, windy conditions 5-7 cm of water may be needed per week wherein arid regions 10 cm may be required. However, the frequency of water application depends on the total supply of available moisture to the roots and the rate of water use, which is influenced by factors such as rainfall, plant age, rooting depth, soil type, crop cover, temperature, humidity and wind. Though brinjal cannot tolerate water logging, timely irrigation is essential especially for fruit set and development.

Drip irrigation has been found beneficial for reducing water use and weed growth as compared to conventional surface irrigation. Trickle irrigation is often used if black plastic mulch is applied. Gibbon in 1973 found that an area irrigated by drip or trickle subsurface irrigation produced bigger and healthier plant, more fruits and higher fresh and dry weights than a similar area surface irrigated. Yields without fertilizers were greater with subsurface system, and optimum fertilizer levels were higher with subsurface than surface irrigation.

Lenz (1970) while studying the influence of irrigation on growth and water consumption concluded that compared with vegetative plants, fruiting egg plants had a smaller total dry weight and smaller water consumption. Anyhow the transpiration was increased by the presence of fruits.

Abdelfattah and Abdel Salam (1972) reported that when plants were watered every one, two or three weeks, the depth of root penetration increased as the interval between irrigation was lengthened. Whilst the extension, the main stem, the number of branches per plant, the average fruit weight and the number and weight of marketable fruits per plot were reduced, and the root penetration was deeper when 100 m³ water/feddan was applied than when 200 m³ water/feddan was applied. While Umrani and Khot (1974) reported that higher yield and moisture use efficiency at 45 mm pan evaporation level. Plastic as ground cover tends to improve the yield as it warms the soil, conserve moisture and controls weeds (Lindgren, 1982). Place plastic mulch in the soil beds before transplanting and make holes where seedlings are transplanted.

1 Irrigation is needed for normal growth and yield of the plant.

2 It is needed for metabolic processes of the plant.

3 To reduce the soil temperature.

4 For easy germination of the seeds from the soil.

5 Irrigation water acts as a medium for transport of nutrients and photosynthates in the plant system.

6 To provide crop insurance against short duration drought.

7 To wash out, dilute salts in the soil.

8 It helps in reducing the hazards of soil piping.

9 To soften tillage pans.

Nutrient Requirement

Brinjal is a long duration crop occupying the land for nearly 6-8 months, requires a good amount of manures and fertilizers for higher yields. Besides, it is a heavy feeding vegetable crop removing large quantities of plant food in one cycle of plant growth. Higher fertility levels and better soil conditions have significant and positive effect on productivity. Flower and fruit production will be adversely effected when the crop is grown under low fertility conditions. However, quantity of manure and fertilizer required for brinjal cultivation depends upon the fertility status of the soil where it is grown. As a general rule 250-300 q/h of farm yard manure should be applied and thoroughly incorporated into the soil at the time of final ploughing. In addition to organic matter, nitrogen, phosphorus and potash should also be applied in the ratio of 100 : 80 : 60 kg/ha. NPK application alone has less response than when applied in combination with organic manures. Half quantity of nitrogen and full quantities of each phosphorus and potash is applied at the time of transplanting. While, remaining quantity of nitrogen is applied either twice or thrice depending upon the soil conditions at 30 days and 45 days after transplanting. Foliar application of 1 per cent urea increases the growth and fruit set. Application of wet cow dung as a band, 10-12 cm away from the plant, followed by earthing up at fortnightly interval during rainy season is a common practice for high production in Kerala. Gnanakumari and Satyanarayana (1971) while studying the effect of N, P and K by applying equal quantities of N, P and K at the rates up to 392kg per hectare on flowering, yield and composition of brinjal, reported that flowering was advanced by all treatments. Plants receiving 224kg per ha. rate flowered after 38 days compared with 65 days in un-fertilized control plots. The highest weight, yield and number of fruits were obtained in response to N, P and K at 280 kg/hectare. Umrani and Khot (1973) reported that highest yield could be obtained with 112 kg N/ha, they further observed that P had no response. Chandrasekaran and George (1973) while applying N, P and K rates up to 125, 200, 100kg and 300kg Spartein (secondary nutrients and trace elements) per hectare, reported that highest N level significantly increased flower production. Though there was slight increase in yields in response to N and P but had no response to Spartin and K.

Aliev (1973) reported that the optimal time for P and N application was before planting and during fruiting respectively. 7.6 kg N, 1.4 kg P and 17.3 kg K per hectare are required to produce one tonne of fruits in CO_2 cultivar, was reported in Tamil Nadu. Apply farm manure and neem cake as basal fertilizers (Sridhar *et al.*, 2002). Watering newly transplanted plants well with sea weed extract or with compost tea will give the seedlings a good start. To prepare your own compost tea, mix 1 part of compost with 6 parts of water. Leave the mixture for one week. Strain and spray on seedlings to control fungal pathogens and prevent infection (Ellis and Bradley, 1996). Forty days after transplanting, side dressing with groundnut cake is recommended. Also at the period, remove 3 nodes at the tips of the plant to improve branching and to increase the number of fruits (Sridhar *et al.*, 2002).

In-order to avoid micronutrient deficiency in brinjal, foliar sprays of B, Mn, Cu and Zn should be applied as they not only enhance the fruit yield but also stimulate photosynthesis and increase sugar, dry matter and vitamin C contents of the fruit. Therefore, foliar application of potassium nitrate (KNO_3 2000 ppm), boron and zinc sulphate (500ppm) increase the yield due to higher photosynthesis and effective translocation of the photosynthates. Singh and Sandhu (1970) compared foliar application of N at 1, 1.5, 2 or 2.5 per cent with soil application of 25, 50, 75 and 100 kg per hectare and found that growth and yield significantly increased by increasing doses of soil applied 'N' up to 75 kg per ha. and the foliar spray at the lowest concentration. Higher concentrations lowered the yield. While Seth and Dhauder (1970) reported that egg plants responded well to 'N' fertilization at 70 kg per ha. in respect of both fruit and seed yield, but there was no significant response to 'P.'

The nutrient requirement for varieties is 100 : 50 : 50 kg NPK/ha and for hybrids it is 200 : 75 : 75 kg/ha

Basal Dressing

Fifty per cent of nitrogen is applied alongwith full dose of phosphorus and potash as basal dressing just before planting as given below.

1. 50 kg N (110 kg of urea) per hectare.

2. 50 kg of phosphorus (313 kg of superphosphate) per hectare.

3. 50 kg of potash (80 kg of muriate of potash) per hectare.

Top Dressing

1. Thirty days after transplanting, the remaining 50 per cent nitrogen is applied in the form of urea 110 kg, as a band application 5–10 cm away from the plants and mixed well with the soil.

2. Immediately, the plants are earthed up and irrigated.

3. NPK at 300 : 50 : 90 kg/ha and 75 × 60 cm spacing was found optimum for PLR1 brinjal.

Annamalai brinjal responded well for ratooning with yield potential of 63 per cent of the main crop within in 100–110 days and with a fertilizer dose of 75 : 25 : 30 kg NPK/ha plus Azopirillum and phosphobacteria each 2 kg/ha.

However, the quantity of nitrogen, phosphorus and potash, on the basis of several years of experiments under All India Coordinated Vegetable Improvement Project, following recommendations have been given (Anon, 2000).

☆ Nitrogen at 150 kg/ha and phosphorus at 100 kg/ha were found optimum at IIHR, Bangalore.

☆ Nitrogen at 100 kg/ha and phosphorus at 60 kg/ha has been recommended for variety H-4 and Pusa Purple Long under Hissar conditions.

☆ Nitrogen at 120 kg/ha has been recommended for Bhubneshwar conditions.

Table 13: Recommended Manuring Schedule for Different States of India

States	N	P_2O_5 Kg/ha	K_2O	FYM (t/ha)
Andhra Pradesh	100	60	60	20
Assam	50	50	50	10
Bihar	160	88	90	20
Gujarat	100	50	50	10-15
Haryana	100	50	25	25
Himachal Pradesh	45	60	40	—
Karnataka	125	100	50	25
Kerala	75	40	25	20-25
Madhya Pradesh	100	60	25	20
Maharashtra	90	40	0	50
Orissa	125	80	110	—
Punjab	125	62	30	25
Tamil Nadu	100	50	30	25
Uttar Pradesh	100	50	50	20
Kharif	100	40	40	25
Rabi	100	50	50	20-22

Half quantity of nitrogen and full quantities of each phosphorus and potash is applied at the time of transplanting. While, remaining quantity of nitrogen may be applied either twice or thrice depending upon the soil conditions at 30 days and 45 days after transplanting.

Deficiency of Major and Minor Elements

Ntirogen

General paling and stunting of the plants. Leaves pale green, older leaves small and uniformally pale green. Start getting bleached from margins inwards until finally entire leaf is bleached to pale white.

Phosphorus

Dirty grayish of young green leaves and premature shedding. Leaves become small and grayish in the beginning, later turning to dirty graying green.

Potash

Interveinal chlorosis of the young leaves followed by yellowing and premature shedding. Retard growth is lesser in comparison with N & P. Leaves normal green but smaller in size with crinkled surface. Small whitish necrotic spots on entire lamina in older leaves.

Calcium

Light green colour of the young leaves alongwith necrotic spotting. Irregular chlorotic lesions on young leaves, which increase in site, coalesced and form bigger patches. Later, these symptoms gradually proceed downward. Calcium deficiency depresses the growth of brinjal, flower and fruit formation. The fruits are small, deformed, reduced in weight, length, width and volume. The biomass, concentration of calcium in leaves and fruits, contents of chlorophylls, sugars and starch are reduced and that of concentration of phenols and nitrate nitrogen increase significantly in calcium deficient brinjal. Low calcium also decrease the activity of peroxidase, ATPase, invertase and α-amylase in brinjal leaves.

Magnesium

Inverted 'V' shaped interveinal chlorosis, more marked in central areas between veins. The fruits are small and may be shed.

Sulphur

Plants deficient in sulphur are small and spindly with short and slender stalks. Retards growth and maturity. The growth of shoot is more affected than root growth.

Boron

Death of the apical bud coupled with browning of roots. Its deficiency disturbes changes in metabolism, increases the accumulation of phenols as well as polyphenol oxidase in cell walls (Hendricks and Van Loon 1990).

Copper

Plants are severely stunted and chlorotic with green spots on the leaves and marked yellowing and browning at the leaf tips.

Iron

Interveinal chlorosis of the young leaves followed by yellowing and premature shedding.

Interculture

It is essential to keep the field free of weeds especially at initial stages of crop growth and is usually done by 2-3 light hoeing or earthing up. This facilitates better aeration to root system and gives support to plants. Application of fluchloralin one week after transplanting seedlings @ 1.5kg a.i/ha as a pre-emergent weedicide, followed by one hand weeding. At 30 days after planting controls most of the weeds. Use of black polythene mulches is also efficient for suppression of weeds and for better growth of plants. In general, weeding, hoeing and earthing up should be done along with fertilization.

Weeding facilitates good aeration and better development of root system. It is a process of eliminating copetition of unwanted plants to the regular crop of brinjal in respect to moisture and nutrition so that crop can be grown more profitably. Weeding also facilitates other operations like irrigation and fertilizer application. In other words, it conserves soil moisture, reduces competition for nutrients and water and maintains purity of seed.

Earthing up and ridges–it is the putting the earth or soil just near the base of the plant. It gives support to the plant when heavy with fruits. Earthing up provides more soil volume for root growth and facilitates uniform spread of moisture during operation of irrigation.

Other Operations

Certain other operations like gap filling, thinning and propping are required as part of inter-cultivation operations.

Likewise thinning is also practiced in direct sown crops, to avoid over crowding and to maintain uniform plant stand.

Growth Regulators

A growth regulator is an organic compound. It can be natural or synthetic. It modifies or controls one or more specific physiological processes within a plant, but the sites of action and production are different.

If the compound is produced within the plant it is called a plant hormone (*e.g.* auxin, which regulates the growth of longitudinal cells involved in bending a stem of a plant one way or the other). Substances applied externally also can bring about modifications such as improved rooting of cuttings, increased rate of ripening, easier separation of fruit from the stem etc. A large number of chemicals tend to increase the yield of certain plants such as corn and sugarcane.

Auxin

These are organic substances which at low concentration (less than 0.001m) promote growth along the longitudinal axis, when applied to shoot of plants freed as far as placing from their own inherent growth promoting substances.

Gibberellins

These substances are having gibbane ring skeleton capable of producing the same physiological responses as gibberellic acid that it must be active in specific gibberellin bioassay. The gibbereins are phytohormones which are active in regulating dormancy, flowering, fruit setting and stimulating germination of seeds and extending growth of shoots.

Cytokinins

These are substances composed of hydrophyllic group of higher specificity (adenine) and one lipophilic group without specificity. The cytokinins form a group of plant hormones having similar effects as those of GA3 in breaking the dormancy of a wide range of seeds and in increased fruit set. These hormones mainly *stimulate cell divisions* and prevent *chlorophyll degration*. Cytokinin stimulate plant growth at critical stages of development leading to increased yield, increased seed or fruit set, larger fruit, increased resistance to environmental stress, earlier maturing and improved crop quality

Absecisic Acid (ABA)

ABA is a naturally occurring sesquiterpene which regulate plant growth and metabolism in various ways and has been detected in nearly all plants. It is involved in abscission of plant organs, induction and vegetative buds, in regulation of fruits ripening and generally in reduction of growth.

Ethylene

It is the only gaseous hydrocarbon hormone which plays an important role in the ripening of fruits, inhibition of root growth, abscission and other growth processes. Unlike the other hormones, ABA and ethylenes are not discovered through any interaction with fungi.

The Role of Plant Growth Regulators in Vegetable Production

The role of plant regulators in various physiological and biochemical processes in plants is well known. Growth regulators are known to affect:

☆ Seed germination.

☆ Seed dormancy

☆ Vegetative growth.

☆ Nodulation

☆ Tuberization

☆ Fruit ripening and yield.

☆ They can also be used for producing polyploidy and male sterility in order to overcome inter specific incompatibility and for producing hybrid seeds.

They act upon the physiological processes within the plant. They help to regulate a wide range of processes including seed germination, root growth, vegetative growth, flowering, fruit growth and development and post harvest preservation in such a way as to improve overall crop production programes.

Effect of Growth Regulators on Brinjal

Nitrobenzene in combination with PGR's produce best results, as it increases flowering, and prevents shedding of flowers in brinjal. There is enhanced flowering and plants take less time to flower. The yield contributing characters like plant height, the increase is about 8-10 per cent. Number of branches per plant increase is 15 to 20 per cent. Number of flowers per cluster, number of fruits per plant and fruit weight increases 20 to 30 per cent, 20–30 per cent and 20–25 per cent respectively. Application of nitrobenzene increases fruit size and the total yield by 35–50 per cent. There is also increase in fruit quality characters like TSS (6–7 per cent), Acidity (7–10 per cent) and Total sugar (5–7.5 per cent).

Application of 2-4,D (2ppm) as whole plant spray at an interval of one week from 60-70 days after transplanting, from the commencement of flowering increases fruit set, early yield and total yield in brinjal. Spraying Mixtalool (long chain C24-C34 aliphatic alcohol) at 4 ppm, 4-6 weeks after transplanting is also effective in increasing yield by 7.1 per cent in F1 hybrid Arka Navneet.

Soaking of seedling roots in NAA at 0.2mg/ha and ascorbic acid at 250ml has been reported to produce highest fruit yield.

According to Jayaram and Neelakandan (2000) plant growth regulators such as IAA, GA3 and Ascorbic acid at

10, 25 and 50 ppm concentrations influenced the development of flowers in brinjal and observed that 25 ppm and 50 ppm treated plants produced significant number of female flowers and same concentrations of IAA and GA3 produced significant number of male flowers. These results highlight the tendency to altered sex of brinjal flowers and produced significant number of female and male flowers by external application of AA, IAA and GA3.

Gavaskar and Anburani (2004) while studying the effect of different concentrations of plant growth regulators such as NAA, 2,4-D, GA3, Kinetin, and Ethrel, observed that GA3 at 200 ppm increased plant height (124.10 cm), increased the leaf number (89.27), number of primary (13.19) and secondary branches (16.79) as compared to control. They further observed that application of GA3 at 15, 30 and 45 days after transplanting was the best for increasing the growth parameters in aubergine.

Aheer *et al.* (1996) studied the effect of three insect growth regulators *viz.*, Match 50 EC, Insegar 25 WP and Atabron 5 EC, @ 500 ml, 500 g and 25 ml and two insecticides *viz.* Polo and Cypermethrin @ 120 ml and 625 ml per hectare, sprayed 4 times at an interval of 15 days, against fruit and shoot borer respectively. It was Isegar that proved significantly more effective against the pest with minimum infestation of 12.44 per cent and and recovered marketable yield of 17458.55kg per hectare.

A study conducted at SKN College Jobner-Rajasthan during 2003 by Meena and Dhaka, revealed that two sprays of GA3 (100 ppm) at 35 and 45 dayts after transplanting recorded 47.54 per cent more yield. Both GA3 and NAA treatments significantly influenced the over all performance of brinjal over control. The treatments comparised of four levels each of GA3 (50, 100, 150 and 200) and NAA (25, 50, 75 and 100 ppm) along with control. The maximum plant height (50.23 cm) and number of branches/plant (15.35) was recorded with spray of 50 ppm NAA. Whereas, the maximum plant spread (1.2007 m²) and fruit size (6.39 cm) was recorded with spray of 150 ppm GA3.

In an another experiment on the effect of growth regulators *viz.* GA3 and NAA on seed yield of brinjal by Patil *et al.* (2008) revealed that GA3 500 ppm significantly increased seed yield (33.96 g) per plant, germination (75.52 per cent), root length (7.81 cm), shoot length (6.86 cm), seedling vigour index, seedling dry weight (14.12 mg compared to NAA 40 ppm and control (30.97 g, 71.89 per cent, 7.45 cm, 859 and 12.39 mg respectively) irrespectively of growth, 4 fruit retension per plant recorded significantly highest germination, root length, shoot length, seedling vigour index and seed weight (83.28 per cent, 8.43 cm, 7.48 cm, 1136 and 16.85 mg) respectively compared to control.

A field experiment was conducted to study the effect of PGR's *viz.*, 40 ppm NAA, 10 ppm GA3, 2 ppm 2,4-D, 300 ppm ethephon, 30 ppm BAP and 5 ppm triacontanol on the morphological characters and yield of Pusa Purple Long and Pusa Purple Cluster at Khajura, Banke district during summer-raining season (Sharma, 2006). The study revealed that PGR's had no significant effect on plant height and stem diameter at the end of crop period and 100 days to flowering. The PPL was earlier to 100 per cent plant flowering, which took 33 days after transplanting. The treatments had no significant influence in fruit number per plant and fruit yield. The interaction effect showed that the PPL did not produce statistically different fruit number per plant with respect to growth regulators, while it had significantly higher fruit yield (17.76 t/ha) at 40 ppm NAA than that at 10 ppm GA3 and 30 ppm BAP. The PPC produced significantly higher fruit number per plant and higher fruit yield (t/ha) at 30 ppm BPA than all other treatments except 5 ppm triacontanol.

Meena and Dhaka carried out an experiment during 1997–98 to determine the most effective and the economical PGR's in brinjal (cv. BR-112) under semi-arid conditions of Rajasthan. The treatments comprised; 50, 100, 150, and 200 ppm GA3; 25, 50, 75 and 100 ppm NAA and control. The plants were sprayed at 35 and 45 days after transplanting. The application of GA3 and NAA increased the yield and net returns compared to control. The cultivation costs was 17,856.96/ha which included labour costs, tractorization, irrigation, manuring and plant protection. Brinjal treated with 100 ppm GA3 gave the highest yield (568.82 q/ha). The gross return of brinjal was 250 q/ha. The highest net profit (Rs.1,07,948.07) was obtained with 100 ppm GA3 followed by 50ppm NAA (Rs.1,02,383.27) compared to control (Rs. 78,493.07). However, 50 ppm NAA was the most economical and gave highest cost: benefit ratio (1:5.60) compared to the control (1: 4.39).

In an experiment, brinjal seeds were soaked separately in three growth regulators *viz.* GA, 2-chloroethyl trimethyl ammonium chloride (CCC) and maleic hydrazide (MH)) at 0, 10, 50 and 100 ppm concentrations respectively. The following results were recorded:

Anthesis

The data so obtained indicated that flowering was hastened by GA which showed a significant difference from CCC, MH and control. It further revealed that at lower concentrations these regulators caused earlier anthesis, while at higher levels opening of flowers was delayed. Wheras, GA treated seeds (10 ppm) caused earliest anthesis compared to all other treatments.

Number of Fruits per Plant

It was GA again that produced the largest number of fruits (50 ppm) followed by CCC at 100 ppm and control. While MH reduced the fruit number.

Fruit Size

Both GA and CCC significantly stimulated fruit length as compared to control, while MH sortened it. With regard to the girth of the fruit, CCC significantly thickened the fruits as compared to MH, GA and control.

Yield per Plant

GA significantly increased the yield per plant as compared to other treatments including MH and control. While MH suppressed the fruit yield. 50 ppm was found to be optimum for increasing the yield which is more true in case of GA. While CCC produced fruits were brighter and glossy in appearance as compared to other treatments. Higher concentrations (100 ppm) of each treatment had inhibitory effect on the fruit yield.

Ascorbic Acid

GA treatment positively influenced ascorbic acid content followed by CCC while MH reduced it. Beyond 50 ppm, ascorbic acid content was reduced. GA at 50 ppm level enhanced early flowering and increased yield. CCC followed GA, while MH had suppressed flowering and reduced the yield.

Chapter 7
Major Diseases of Brinjal

Table 14: Fungal Diseases

Diseases	Causal organism (s)
Damping off	*Phythium* spp.
	Phytophthora spp.
	Rhizoctonia spp.
Phomopsis blight	*Phomopsis vezans*
	Alternaria spp.
	Cercospora spp.
Wilt	*Verticillium* spp.
	Fusarium solani
Southern blight	*Sclerotium rolfsi*
	Macrophomina phaseolina
Bacterial wilt	*Pseudomonas solanacearum* E.F.sm.
Anthracnose Fruit rot	*Collectotrichum* sp.
Alternia blight	*Alternaria solani*
Little leaf	*Mycoplasma*
Root Knot of Brinjal	

Fungal diseases are problem in many crops. They can be recognized by several damage symptoms. A single fungal infection is able to cause a combination of these damage symptoms, that can differ per crop.

Biology

Rotting and Wilting

These are often caused by soil pathogens infecting plants, via the roots. Plant strengtheners like TRIANUM-P and TRIANUM-G reduces the chance of infection by soil pathogens. These products do not control fungi, but form a protective layer around the roots. The fungus Trichoderma harzianum strain T-22 in TRIANUM-P and TRIANUM–G put the soil fungi at a disadvantage by feeding on nutrient surplus and by spatial competition. TRIANUM–P and TRIANUM–G are currently approved in Spain and Britain.

Damage Symptoms

Wilting Tissue

Seedlings wilt and usually die off. Leaves and stems of large plants wilt, discolour and whiter.

Dying Tissue (Nicrosis)

Visible as small, yellow brown coloured spots, that can increase in diameter or continuously extends over the tissue. Tissue can get discoloured over larger areas. Spores, visible as powder or "fluff," can be formed on these spots. Leaves and stems are covered by mycelium, usually visible as a thin, feltlike, white till dark gray layer extending over the plant tissue. Development of light orange, yellow or brown discolouring spots/dots, on leaf or stem surface (rust). Spores break through the leaf or stem surface after some time. Visible as small, coloured bulges covered with a similarly coloured fine powder. Deformation of leaves, stem, flowers and or fruits. Root and/or stem rot, visible as rotted plant parts. The tissue usually colours brown or black. Mycelium and spores can become visible as bluff or powder. Tissue get slimy. Plants above usually die off.

Verticillium Wilt

Caused by the fungus *verticillium* spp. It is prevalent and destructive disease. It causes stunting of plants and interveinal yellowing, wilting and drying of leaves. Plants usually survive in this condition, but a few may die. Woody tissue in the lower stem is discoloured brown. The disease is caused by the same soil borne fungus that causes wilt of tomatoes, potatoes, strawberries, and brambles. The fungus is able to persist in soil for many years. Presence of root rot or root lesion nematodes may increase disease severity.

Damping Off (*Pythium* spp., *Phytophthora* spp. and *Rhizoctonia* spp.)

This is a major nursery disease. The fungus usually starts on germinating seeds, spreading to the hypocotyls, basal stem and developing tap root. The affected seedlings are pale green and brownish lesions found at the basal portion of the stem. Lesion girdles the stem, later extending upwards and downwards. The affected tissue rot and the seedling collapses. The disease is very severe in the nursery and is soil borne. In summer damping off is mainly caused by *Rhizoctonia solanii* and the same can be controlled by seed treatment with Carbandazim @ 2g/kg seed followed by nursery dressing with Carbandazim 0.1 per cent.

Control

☆ Sterilization of the nursery soil before sowing.

☆ Treat the seed with Bavistan (0.1 per cent) before sowing.

☆ Hot water treatment at 51.7°C for 30 minutes of seeds is also effective in controlling the disease.

☆ Sow seeds as loose/thin as possible on raised beds.

☆ Adequate drainage and drenching nursery bed with Bavistin (0.1 per cent) controls the disease effectively.

☆ Treat seeds with Captaf @ 2 g/kg of seed followed by drenching the soil with 0.2 per cent Captaf.

☆ Seed treatment with Trichoderma virde@ 4 g/kg of seeds followed by soil application @ 50 mg/5 g f.y.m/m² can also control damping off effectively.

Powdery Mildew [*Leveillula tauica* (Lav.)]

In the beginning, on the lower surface of leaves, the fungus forms dirty white irregular powdery spots, which in due course, coalesce with one another and cover the entire lower surface also. As a result of powdery mildew infection, the leaves fall prematurely.

Control

Spray Karathane 0.1 per cent or sulfex 0.2 per cent.

Southern Blight

Caused by the fungus *Sclerotium rolfsii*, is characterized by softening of crown and external root tissue. Fungus mold and tiny brown sclerotia (fungus reproductive structures) grow over the base of the stem and near by soil.

Symptoms

☆ The lower portion of the stem is affected from the soil borne inoculum (sclerotia).

☆ Decortication is the main symptom.

☆ Exposure and necrosis of underlying tissues may lead to collapse of the plant.

☆ Near the ground surface on the stem may be seen the mycelia and sclerotia.

☆ Lack of plant vigour, accumulation of water around the stem, and mechanical injuries help in development of this disease.

Control

☆ Seed treatment with 4 g of Trichoderma viride formulation per kg seed will help in reducing the disease.

☆ Spraying with Mancozeb @ 2g/Litre of water.

☆ Collection and destruction of diseased parts and portions of the plant.

Phomopsis Blight

(*Phomopsis vexans* Harter) is caused by the fungus *phomopsis vexans*, affects all above-ground plant parts at all stages of development. The fungus is both seed and soil borne. Spots generally appear first on seedling stems or leaves. Spots may girdle seedling and kill the seedlings. Leaf spots are clearly defined, circular, up to 2.5 cm in diameter and brown to grown to gray with a narrow dark brown margin. In time the center of the spot becomes gray and black pycnidia (fungus reproductive structures that appear as small specks) develop in this area. Affected leaves may turn yellow and die. Fruit spots are similar to those on leaves but are much larger, affected fruit are first soft and watery but later may become black and mummified. Phomopsis persists on and in seed and over winters in residue from diseased plants. It is spread by splashing water. Disease is promoted by wet weather and high temperature. It subsists between crop seasons on infested plant debris the organism apparantly dessiminates locally as water borne pyenospores. Wet weather conditions and high temperature (30–32°C) favours disease actively.

Control

☆ Use of disease free seeds.

☆ Blemish free fruits from healthy plants should be selected for seed.

☆ Fruit should be dipped for ten minutes in mercuric chloride solution (25g/35 lit. water).

☆ Growing of resistant varieties/hybrids.

☆ Hot water treatments of seeds at 50°C for 30 minutes.

☆ Follow long duration crop rotation and destruction of infested plant material on the field are important cultural practices to reduce the disease.

☆ Spraying at regular intervals in the nursery and in the main field with Bordeaux mixture (4:4:50) or other chemicals

☆ Adoption of suitable sanitary measures.

☆ Treat the seeds with Agrosan–GN or Thiram (2.5g/ kg seed) or Bavistan (carbendadazim) before sowing (Kushal and Sudha 1995).

☆ Sterilize soil before sowing the treated (0.1 per cent Bavistin) seed.

Alternia Blight caused by the fungus *Alternaria solani,* occurs on egg plant as well as on tomato. Scattered dark brown spots on leaf lets are leathery showing concentric rings and range up to 2.5 to 7.5 cm in diameter, sometimes may cover large areas of the leaf blade. When spots are numerous, leaves die prematurely and drop. Fruit also may be spoted, infested fruits turn yellow and drop off prematurely.

Control

Spray with Bordeaux mixture 5:5:50 or Zineb 0.25 per cent.

Leaf Spot (*Alternaria* spp. and *Cercospora melongenae* Chupp.)

Alternaria spp. produce leaf spots with concentric rings. Spots are irregular, on severe infection, leaves drop off. Fruits become yellow and drop. The pathogen survives in the soil in diseased plant materials. The leaf spots are characterized by chlorotic lesions, angular to irregular in shape, later turn to grayish-brown with profuse sporulation at the centre of the spot. Severely infested leaves drop off prematurely, resulting in reduced fruit yield.

Control

☆ Collect leaves and destroy.

☆ Effective control is to spray Bavistan (0.1 per cent).

☆ Pant Samarat variety is resistant to both the leaf spots.

☆ Regular spray of Dithane-D-78 or Dithane-M-45 or Fytolan or Blitox at 0.2 per cent.

☆ Spraying 1 per cent Bordeaux mixture or 2 g Copper oxychloride or 2.5 g Zineb per litre of water effectively controls leaf spot.

☆ Grow brinjal varieties like Junagadh Long, P-8, Manjari Gota, PPC, H-4, Pant brinjal-7. Pant Rituraj.

Bacterial Wilt

It is caused by *Pseudomonas solanacearum*. It is a major disease of brinjal, specially in areas where the bacterium has become established in the soil due to favourable temperature and moisture. The characteristic symptoms of the disease are wilting, stunting, yellowing of the foliage and finally the entire plant collapses. The lower leaves may drop first before wilting occurs. The vascular system becomes brown and if a segment of the lower stem is cut and squeezed bacterial ooze pours out. The pathogen is mostly confined to vascular regions in advance case, it may invade the cortex and pit and cause yellow brown. If humid weather continues and the temperature is high there is sudden drooping of the leaves without yellowing, and rotting of the stem from any point. The roots may appear healthy and well developed but there is brown discolouration inside.

Since it is a soil borne disease, pathogen persists for longer periods in the infected soils, may enter through wounds caused during transplanting, cultural operations or nematode invasion. The organism first moves into the large xylem vessels. When a single lateral bundle is invaded drooping of leaves is common but if all bundles are invaded the plant wilts.

Symptoms

☆ There is wilting, stunting, yellowing of the foliage and finally collapse of the entire plant are the characteristic symptoms of the disease.

☆ Lower leaves may droop first before wilting occurs.

☆ The vascular system becomes brown.

☆ Bacterial ooze comes out from the affected parts.

☆ Plant shows wilting symptoms at noon-time will recover at nights, but die soon.

Control

☆ Crop rotation can reduce the severity of the disease. That brinjal crop should be rotated with cereal crops or cruciferous vegetable such as cauliflower or cabbage. Superphosphate increases disease while nitrogen reduces it.

☆ Pant Samrat variety is tolerant.

☆ Field should be kept clean and effected parts are to be collected and burnt.

☆ Remove weeds like *phyllanthes neruri*.

☆ Spray Copper fungicides to control the disease. (2 per cent Bordeaux mixture).

☆ The disease is more prevalent in the presence of Root Knot Nematodees, so control of these nematodes will suppress the disease spread.

☆ Treat the nursery soil with formalin 4 per cent (100 ml of 40 per cent formaldehyde in 900 ml of water) for 2 days before sowing of seeds in the nursery bed.

☆ Follow long term crop rotation including Maize, Jowar, Soyabean.

☆ Apply neem cake or groundnut cake.

Fusarium Wilt (*Fusarium oxysporum f sp. lycopersici* (Sacc.):

The disease is characterized by yellowing and wilting of leaves and finally the entire plant wilts and dies prematurely. Stem tissue often is discoloured thoughout the plant. Vascular browning takes place in the root.

Control

☆ Follow crop rotation

☆ Use of resistant cultivars.

☆ Drench the soil with Bavistan 0.1 per cent or Belate 0.2 per cent.

Fruit Rot (*Phytophthora nicotianae* B. de Haan var. *nicotianae*):

The disease first appears as small water soaked lesions on the fruit, which later enlarges in size considerably. Skins of infested fruit turns brown and develop white cottny growth. Humid weather conditions favour disease development. In order to minimize infestation, good drainage conditions should be maintained in the field.

Control

☆ Clean cultivation, proper drainage are important to minimize its infection.

☆ Staking of the plants, removing foliage and fruits up to 15–30 cm from ground level has been found to control the disease.

☆ Mulching followed by spray with Difoltan 0.3 per cent twice at 10–15 days intervals is recommended.

Root Knot of Brinjal

The disease is caused by nematodes, is very extensive especially in tropical and sub-tropical climates. The causal organisms are *Meloidogyne incognita* and *Meloidogyne javanica*. If the infection is early and severe, there is unthrifty development and stunted growth. The leaves are yellowish-green to yellow, tend to droop followed by wilting. Sometimes there is scorching of the leaf from the margin inward. The most characteristic symptom of the disease is the formation of knots or galls on the root system. The main root and the laterals bear spherical to elongated galls which vary in size from very small to large. In advanced stages the galled tissues decay and are invaded by other pathogenic and saprophytic organisms. A large number of females of the nematode are present in the root galls. These survive in the soil in the root debris.

Control

☆ Phorate at the rate of 25kg/ha gives almost complete control of root knot when applied to soil immediately or seven days before transplanting.

☆ Deeper summer ploughing and allowed to dry.

☆ Application of neem (margosa) cake at the rate of 25 q/ha or saw dust at the rate of 25 q/ha followed by application of nitrogen through urea (120 kg/n per ha) has given very good results.

☆ Use resistant varieties like Banaras Giant, Black Beauty and Manjari

Brinjal Little Leaf (Mycoplasma)

Affected plants show the formation of very small leaves and deformation of flower parts. The disease is transmitted in nature by leafhopper (*Hishimonus phycitis*). The early symptoms are seen in the young leaves near the tip of the plant. These leaves are light yellow and thinner than healthy leaves of the same age. The flower buds take on upright position instead of pendulous. Simultaneously or little later, the normally dormant axillary buds sprout. As the disease advances, all the newly produced flowers becomes phylloid, leaves become progressively smaller, the stimulated axillary buds grow out into axillary branches.Two to three months after infection, the diseased plants loose its resemblance to a healthy brinjal because of excessive crowding of short branches and proliferation of smaller leaves which gives "rosette" appearance to the plant. Flowers become virescent and sterile.

Control

☆ The infested plants should be rouged out as soon as they appear.

☆ Apply Phorate at 1kg per hectare in the soil of nursery bed.

☆ Dip seedlings in 0.2 per cent Carbofuran 75 per cent WP for 24 hours before transplanting.

☆ Dip the roots of seedlings at the time of transplanting with 1000 ppm of tetracycline.

☆ Spray crop 4-5 times at an interval of 10 days with methyl parathion (0.05 per cent) or dimethoate 0.02 per cent or monocrotophos (0.05 per cent).

☆ Spray rogor at 0.05 per cent at 10 -15 days interval till fruit set.

Brinjal Mosaic Caused by Tobacco Mosaic Virus

This is a virus disease, is seed borne and transmissible.

The virus is highly contagious and transmitted by sap, contamination through implements, soil debris, hand of the working labour and contimanated clothes. The important symptom is conspicuous mosaic molting and yellowing of leaves followed by blisters as the diseases advances. The severely infested leaves become small and misshapen. The infested plants have stunted growth and some times show concentric rings on the leaf lamina. Vigour of the plant is severely reduced and plants wither and do not bear fruits or may bear one or two fruits. The infested plants bear less number of flowers and fruits. The fruits produced show patches of light and dark areas on the surface and remain small and deformed.

Symptoms

Mosaic moltting of leaves and stunting of plants are the characteristic symptoms of potato virus Y Mosaic symptoms are mild in early stages but later become severe. Infected leaves are deformed, small and leathery. Very few fruits are produced on infected plants. Leaves also develop blisters in advanced cases. Severely infected leaves become small and misshapen. Plants infested early remain stunted.

Control

☆ Remove weeds near seed bed or the field.

☆ Before working in the field, hands should be washed with soap.

☆ Planting of pepper, tobacco, tomatoes, cucumber etc. should be avoided near the brinjal seeds beds as for as possible.

☆ Rough out diseased plants.

☆ Collect seeds from only mosaic free plants.

☆ Spray insecticides like Dimethoate 2ml/litre or Metasystox 1 ml/litre of water to control the insect vector.

☆ Prohibit smoking or chewing of tobacco who are handling brinjal seedlings.

☆ Use virus tolerant varieties like PPC.

Alternaria Leaf Spot

The two species of *Alternaria* occur commonly, causing the characteristic leaf spots with concentric rings. The spots are mostly irregular, 4–8 mm in diameter and may coalesce to large areas of the leaf blade. Severely affected leaves may drop off. *A. melongenae* also infects the fruits causing large deep-seated spots. The infected fruits turn yellow and drop off prematurely.

Control

Spraying 1 per cent Bordeaux mixture or 2 g Copper oxychloride or 2.5g Zineb per litre of water effectively controls leaf spots.

Chapter 8
Major Pests of Brinjal

Brinjal should not be planted after tomatoes, pepper, potato or other solanaceous crops to prevent a recurrence of the same pests and disease pathogens. Rotate egg plants with other crops. Verticillium wilt is controlled by crop rotation and planting crops that are not susceptible to wilt (Lindgren, 1982). Planting egg plant after rice reduces the incidence of bacterial wilt and nematodes. Proper seed selection and treatment and planting in well drained soil reduces seed rot and damping (Chen; Li, Kalb, 2001). Row or the mesh covers also help reduce pest attack. However, during the reproductive stage, hand pollination of flowers is necessary (Lindgren, 1982).

Aphids, Beetles, Butter flies and Moths, Leaf miners, Mealy bugs, Sciarid flies, Spider mites, Thrips and White flies.

Table 15

	Pests
Growth stages	
Seeds	
Sown seeds	Ants
Seedling Stage	
Stem	Cut worm
Leaves	Aphids
	Colorado potato beetle
	Cut worm
	Egg plant flea beetle
	Spider mites
	Stink bug
	White fly
Vegetative stage	
Stem	Egg plant fruit and stem borer
Leaves	Aphids
	Colorado potato beetle
	Cutworm
	Egg plant flea beetle
	Spider mites
	Stink bug
	White fly
Reproductive stage	
Fruit	Fruit & shoot borer
Maturation stage	
Fruit	Fruit & shoot borer

A. Chewing Insect that Feed on above Ground Plant Parts

Beetles

Chewing insects with hard, shell like fore wings which meet in a straight line down the middle of the back.

1. *Colorado Potato beetle*: Yellowish-brown, oval, convex body 9 to 14 mm long, black spots on pronotum (area behind head); five black strips on each wing cover; feeds on terminal buds and may consume entire leaf.

2. *Flea beetles*: Tiny, dark coloured, oval beetles with thickened, jumping hind wings; antennae ½ to 2/3 the length of the body, leave tiny " shoot holes" in the foliage.

3. *Egg plant flea beetles*: Black, slightly hairy body about 2mm long with black legs.

4. *Potato flea beetle*: Brownish–black to black, body about 2.5 mm long, much less hairy than egg plant flea beetle.

5. *Colorado Potato beetle larva*: Soft-bodied chewing insect up to 15 mm long, red, yellowish-red or orange body with black head, three pairs of black legs and two rows black spots on each side of the body, one pair fleshy prolegs on last abdominal segment.

B. Pests with Needle-like Mouth Parts which Feed on Above Ground Plant Parts and Extract Plant Juices

1. *Egg plant lace bug:* Herds of grayish to light brown adults with spiny yellow nymphs; adult 4 mm long, 2 mm wide with 2 pairs of lace like wings and a hood like projection just behind (or over) the head; feeds in colonies on under side of leaves; above, infested leaves have roughly circular discoloured areas.

2. *Two spotted spider mite:* Tiny, almost microscopic pale to dark green pest with 2 or 4 darkly coloured

spots; adult and nymph 8 legged. Larva 6-legged; adult female, oval, 0.3 to 0.5 mm, male more diamond shaped; feeds on under sides of leaves; infested foliage with silvery or pale yellow stipples, leaves eventually pale and dry up; silken weds common on under sides of leaves.

C. Soft-bodied, Root Feeding Pests

1. *Flea beetle larvae:* Several species of cylindrical, whitish larvae with brown heads, up to 4–5 mm long with 3 pairs of legs near the head and a pair of anal prolegs.

Shoot and Fruit Borer (*Leucinodes arbonalis* Guen)

Nevertheless, the most serious pest of brinjal is without doubt the brinjal shoot and fruit borer, a Lepidopterous pest whose larve are well protected from pesticides and natural enemies, once they have entered the fruit or young shoots.

It is a very destructive pest of brinjal, widely distributed throughout the country. Apart from Indian sub-continent it has been reported from South Africa, Congo and Malaysia also. The pest is active throughout the year at places having moderate climate but its activity is adversely effected by severe cold. The damage by this pest starts right from the transplanting of the crop and continues till harvest of the fruits or till the crop is there in the field. Eggs are laid singly on the ventral surface of the leaves, shoots, flower buds and occasionally on fruits also. In young plants the caterpillars bore into the petioles and midribs of large leaves and young tender shoots, close the entry point with their excreta and feed within. As a result the effected leaves dry and drop off while, in case of shoots growing point is killed. Appearance of wilted dropping shoots in a crop is the typical symptom indicating damaged by this pest the effected shoots ultimately wither and dry up. At later stage of the growth of the plant the caterpillars bore into flower buds and fruits entering from under the calyx leaving no visible sign of infection while the larva feed inside. The damaged flower buds are shed without blossoming whereas the fruits may show circular exit holes. Affected fruits become partially unfit for consumption.

Control

★ Rotation of crops-continuous cropping of brinjal crop favours heavy infestation.

★ Prompt destruction of the damaged shoots on community wide basis is essential.

★ Two to three sprays of 0.05 per cent endosulfan at fortnightly intervals have been found effective in controlling the pest population.

★ 3 sprays of Carbaryl 50 per cent w.p. 2.5g or Monocrotophos @ 1.25 ml per lit of water. A safe period of 10 days should be maintained between spraying and harvest.

★ Spray insecticide *viz.* Cypermethrin@ 30g a.i/ha at 10 days intervals or deltamethrin @10g a.i./ha.

★ Soil application of Carbofuran 40, 55, and 70 days after transplanting controls the incidence of this pest.

★ Cultivars resistant to Shoot and fruit borer should be preferred *viz.*, Pusa Kranti and Pant Samrat.

☆ Effected plant parts should be removed alongwith larvae and destroyed by crushing or by immersing in insecticide solution.

☆ Spray Fenvalerate 0.005 per cent at fortnightly interval 21 days after transplanting onwards (Bagle, 1993).

Brinjal Stem Borer (*Euzophera particella*)

Brinjal Stem Borer (Euzophera particella) is a another destructive pest of brinjal which also attacks chilli, tomato and at times even tomato. It is widely distributed over Indian sub-continent. Eggs are laid singly or in clusters on young leaves, petiole or even on tender shoots. Soon after hatching the caterpillars bore into stem and move downwards. The attacked plants wither and wilt, the growth is stunted and the fruiting is adversely effected.

Control

☆ Collection and destruction of the infested leaves alongwith insects in the initial stages help to minimize the infestation.

☆ Spray 0.1 per cent Carbaryl or 0.05 per cent malathion.

Brinjal Leaf Roller

Antoba (*Eublemma olivacea*) is a common foliage pest of brinjal. Besides brinjal it also attacks other wild solanaceous plants. Eggs are laid on leaves. On hatching the caterpillars fold leaves and feed within by scraping the green matter. As a result of this damage the folded leaves wither and dry up. Pupation takes place within the folded leaves. The affected plants wither, growth is stunted and fruiting capacity is adversely affected.

Control

☆ Collect and destroy rolled leaves with larvae and pupae inside.

☆ Spray the crop with 0.05 per cent Endosulfan or 0.2 per cent Carbaryl if the infestation is severe.

Leaf Feeding Beetles

Popularly known as hadda beetles. In this are a group of coccinellid beetle of the genus Epilachna and are commonly known as Epilachna beetle.

Epilachna Beetle

These phytophagous and harmful beetle of the genus Coccinella which are predatory beetle and are beneficial insect. A few species of Epilachna beetle have been reported as quite serious pest of a number of vegetable crops from different parts of world.

Eggs are laid in masses on ventral surface of the leaves. A single female lays about 120–180 eggs in 4–8 batches. Both grubs and adults feed voraciously, by scraping chlorophyll from epidermal layer of leaves and tender parts of the egg plant. As a result of their feeding scletonized patches develop on the leaves which gradually dry away. They cause serious damage during their larval stage and when they appear in large numbers. Eggs are elongated, cigar shaped and yellow in colour. Yellowish grubs have spines all over their bodies. Pupation takes place on the leaves and the pupae are hemispherical in shape. Incubation period varies from 2–4 days, grub stage lasts for 12–18 days and pupal period varies from 3–6 days. The entire life-cycle is usually completed in 18–25 days during summer and may extend upto 50 days in winter.

Control

 ☆ Collecting and destroying the effected leaves and egg clusters.

 ☆ Spray with 0.1 per cent Endosulfan or Malathion 0.16 per cent @ 3 ml per water of Methyl parathion 0.03 per cent @ 1 ml per lit. of water @ 1 ml per lit of water.

 ☆ Dust with Chloropyrophos at 20 kg/ha when attack is before fruiting.

 ☆ Plant tolerant cultivars like Arka Shirish, Shankar, Vijai.

Thrips

They attack egg plants mostly during the dry season and cause boring of the leaves, especially on the lower leaf surface and the scarring fruit.

Control

Spray with Dimethoate (Rogor) @ 1ml/litre or Monocrotophos (Nuvacron) @1.5ml/litre or Malathion @1ml/litre.

Jassids (*Amrasca biguttula*)

These insects (both nymphs and adults) suck the sap from the lower surface of the leaves and inject toxic saliva into the plant tissues. The infested leaf curl upward along the margins which may turn yellowish and show, burnt up patches. They also transmit mycoplasma disease like little leaf and virus disease like mosaic. Fruit setting is adversely affected by the infection.

Control

☆ Pre-transplant seedling dip in 0.2 per cent carbofuran 75 per cent WP for 24 hours.

☆ Spray insecticide like cypermethrin (30g a.i./ha) or deltamethrin (10g a.i./ha) or endosulfan (700g a.i./ha) at 15 days interval, starting from 10 days after transplanting.

☆ Apply phorate or aldicarp granules at 1.0 kg/ha 15 days after transplanting followed by 3 fortnightly sprays of carbaryl (0.2 per cent) after fruit set.

☆ Avoid crowding of the seedlings in nursery beds.

Leaf Hoppers

They feed mainly on the under side of egg plant leaves, causing yellow patches on the foliage. Certain species also transmit mycoplasma like disease, such as little leaf disease. Fruit setting is adversely affected by the infection.

Brinjal Mealy Bug (*Centrococcus insolitus*)

There is stunted growth of the plant. The plants appear as though covered with white wash.

Control

Malathion 0.15@ 3ml per lit of water or Monocrotophaos @ 0.4 per cent 1.25 ml/of water.

Brinjal Mite (*Tetranychus telarius*)

Leaves present a blotching appearance, becomes whitish and brown patches develop.

Control

☆ Spray Wettable Sulphur @ 3-5g/lit or Dicofol @

2.7 ml/lit of water or dust Sulphur @ 20 to 25 kg/ ha or spray with malathion or Aramite (0.02 per cent).

☆ Spray sevisulf (a mixture of carbaryl and sulphur) 0.1 per cent.

Aphids (*Aphis gossypii* Glov)

The most common and injurious species of aphid are *Myzus persicae* and *Aphis gossipii*. Aphids or plant lice are polyphagous pests and have a wide range of host plants and are cosmopolitan in distribution. The aphid is small and soft, found in colonies of hundreds on tender shoots. They are sucking insects and suck the sap of leaves, both winged and wingless and are blackish, brownish or green in colour. Both nymphs and adults suck the cell sap from leaves and the tender apical shoots. Heavy puncturing and drainage of sap by large number of nymphs and adults reduce the vigour of the plant. The effected parts turn yellow, wilt, get deformed and fall prematurely. This finally effects adversely the growth of the plant. They also secrete honey dew on which sooty mould grows.

Black sooty mold develops rapidly on the sugary excretions of the aphid. This sooty mold covers the plants, thereby adversely reducing the photosynthetic activity, thus weakening the plant. The infested parts become weak, pale and stunted in growth resulting reduced fruit size. Ants also visit these plants for honey. These pests may also transmit virus diseases and bacterial and fungal organisms can gain entry to the plant tissues through the feeding wounds. Aphids occur in the cool dry season.

Control

☆ 2–3 sprays at 10–12 days interval with 0.05 per cent Dimethoate, Endosulfan or Monocrotophos or spraying of Malathion 0.1 per cent or Metasystox 0.05 per cent or Nuvacron 0.05 per cent can effectively control these aphids.

☆ Apply carbofuran granules @1500 g.a.i during transplanting followed by two to three sprays of cypermenthrin @ 30 g.a.i./ha at 10–15 days interval.

Mites

This pest suck the sap from the lower surface of leaves, where they remain in colonies covered with silky webs. In advance stage of infestation whole field looks yellowish and burning effects. Warm and dry season favours multiplication of jassids.

Control

☆ Spray of neem oil garlic solution in initial stage of attack is effective for control of mites.

☆ Under severe infestation, spray Kelthane (0.03 per cent) or metasystox (0.03 per cent).

☆ Spray with Aramite or Malathion (0.02 per cent) or Dicofol 0.2 per cent.

☆ Dust with fine Sulphur 20–25 kg/ha or spray 0.05 per cent wettable sulphur.

☆ Spray sevisulf (a mixture of carbaryl and sulphur) 0.1 per cent.

Root Knot Nematodes (*Meloidogyne* spp.)

It is one of the most destructive parasite of brinjal plant.

Table 16: Key to Problems

Symptoms	Causes	Control
Blossoms drop; No fruit develops	Poor pollination due to unfavourable temperatures.	Be patient, fruit will set when the temperatures becomes more favourable.
Plants wilt, bottom leaves may turn yellow	*Dry soil*	*Supply water*
	Verticillium wilt (fungal disease)	Rotate. remove old plant debris, do not plant after tomatoes, potatoes, brambles or strawberries
	Water logged soil	Improve drainage
	Root knot nematode	Check roots for knots, rotate, soil pasteurization
Circular or irregular brown spots on leaves and fruits/or fruits	Fungal or bacterial disease	Submit sample for diagnosis
Leaves riddle with tiny holes	Flea beetles	Use insecticide
Large holes in leaves, caterpillars present	Tomato horn worm	Handpick, use insecticide or *B. thuringiensis*

Table 17: Cultivars Showing Resistance/Tolerance Against Few Diseases & Pests

Bacterial Wilt	Arka Nidhi, Arka Keshav, Arka Neelkanth, ARU-2C, Pant Samrat, Swarna Shree, Manjari, PPC, Swarn Prathibha, BWR-12, Swarna Shakti, Swarna Shaymali, Swarna Shobha, Swarna Ajay, SM-6-7, SM-6-6, SM141, Cipaye, Aroman, Gowok, Surya, L-7& 19, PantSinampire, Taiwan Naga, Rituraj, Comuy, Long green, Muktakshi, PPR, Annamali, Utkal Tarini, JC-1 & JC-2
Little Leaf	S. integrifolium, (H-6, 7, 9, 10)P P R, PPC, Bhagmati, H-8, S-212-1 & S-252-1.
Mosaic	PPC, PPR, Surati, S212-1
Phomopsis Rot	Pusa Bairav, Swarna Shakti, Muktakesi, Bargan, White Gih, Pant Samrat, Pant Rituraj, Florida Market, Florida beauty, PPC, Line-119-12-2-264-1-2, EC-384565, PB-30, BC-1, JC-2, KS-352, IC-90922, 90047, 90970.
Early Blight	PPC
Anthrac nose	Ping Tung, Long Red.
Collar Rot	Junagadh Long
Epilachna beetle	Arka Shirish, Shankar, Vijai
Shoot & Fruit borer	Doli-5, Punjab Jamuni Gola, Arbhjit, Punjab Barsati, Punjab sadabhar, SM-17-4, P & G-129-5, PPR, AR2C, Kalyanpur-2, Punjab Chamkila, Swarnashree, Aushey, Junagadh Long, Pant Samrat, H-4, Pusa Kranti, Neelam, Punjal Neelam, H-165, H-407, H-408, PPC, Annamala, Bhagyamati, AM-62
Jassid	Black Beauty, Ramnagar Giant, Vijay Hybrid, Pant Smrat, Mangari Gota

Contd...

Nematodes	PPC, Pant Rituraj, Arka Nidhi, Arka Kesar, Arka Neelkanth.
Fusarium wilt	S.incanum
Meloidogyne spp. resistant	*S.sisymbriifolium*
M.incognita resistant	Black Beauty & Vijaya.
M.javanica resistant	Banta & Black Beauty.
Frost tolerance	*S.incanum.*
Drought tolerance	S. microcarpon.
High temperature	R-34
Multi-resistant (root knot nematode + phomopsis fruit rot+ verticillium wilt)	Floreida Market.

Brinjal is highly susceptible to the nematode, as plant becomes stunted, leaves show chlorotic symptoms and fruit is also adversely affected.

Control

Use nematicides and resistant hybrids.

The Major Constraints in Research

1. In sufficient germplasm available and the need to augment indigenous and exotic collections.
2. Lack of new high yielding varieties/hybrids, carrying high degree of resistance to diseases (bacterial wilt and little leaf) and pests *viz* fruit and shoot borer.
3. Standardisation of agro-techniques for growing exotic as well as indigenous collections.
4. Breeding for nutritional, processing and export qualities.
5. Studies on the insecticidal residues.
6. Development of varieties with longer shelf life.
7. Absence of high yielding hybrids suited to all climatic zones is a major constraint.
8. Absence of certified classes of seed.
9. Brinjal fruits are susceptible to several fruit diseases both under field and as well as storage conditions, besides to those foliage and vascular diseases.

Chapter 9
Harvesting and Yield

==========

It is an operation of cutting, picking, plucking, digging or combination of these for removing the useful part or economic end product, part from the plant. Crop can be harvested at physiological maturity or at harvest maturity. Crop is considered to be at physiological maturity, when the translocation of photosynthates are stopped to economic part. If the crop is harvested early, the produce contains high moisture and more immature grains.

A fruit or vegetable is a living and respiring edible tissue. Development starts with the formation of the edible part, such as the setting of a fruit, emergence of a seedling, the swelling of a root, tuber or bulb or the elongation of a stalk or petiole. Development occurs largely before harvest and includes prematuration and part of maturation. The maturation in fruits and vegetables reaches when the edible portion has attained some desirable conditions. The maturity in the horticultural crops may be defined as the stage of development, when a plant or plant part possesses the prerequisites for utilization by consumers for a particular purpose. However, it may be determined (i) visual means like size, colour, shoulder growth etc, (ii) physical means like firmness, specific gravity etc. (iii) computation as days after fruit set or growth and heat units, (iv). physiological methods like respiration and (v) chemical methods like TSS, acid test, vitamins, starch etc. In case of brinjal, it has been observed that 19 days old fruits are optimally mature to harvest. The surface of the fruit is bright and glossy in appearance. The fruit is edible ripe when it is quarter grown and becomes fully ripe when colour changes to yellow or brown and flesh turns dry and tough.

Brinjals should be harvested at immature stage or as soon as it attains a good size and colour or before the skins turn dull, are small and glossy *i.e.* have not lost culinary qualities.The attractive bright, glossy appearance having freshness and optimum size of fruit are qualities for good market. The dull appearance indicates they are over ripe and will be more prone to bitterness and indicates over maturity and loss of quality. Pressing the thumb against the side of the fruit can test the maturity of the fruit. If pressed portion springs back to its original shape, the fruit is too immature.The harvesting starts from the 50th day onwards and continues for 50 days in the first phase and a second phase can be obtained after 20-25 days if adequate nutrition and irrigation are provided. The crop can be removed after 110 days if the 2nd flush is not desired, otherwise it can be retained for 150–160 days. Fruits should be harvested in the afternoon in order to avoid sunscald.

Therefore, fruits are sprinkled with water after harvesting to keep them fresh. Usually fruits are harvested alongwith its stalk with a slight twist by hand. Some cultivars like BR112 and NDB have soft joints and are easy to harvest. In some varieties, a sharp knife is also used for harvesting fruits along with fleshy calyx and a portion of fruit stalk. While harvesting care should be taken to avoid injury to the bunches. The size of the fruit reduces during summer. The calyx is purple blue and is soft and edible too. During summer the calyx tends to become green. A heavier crop can be produced if the fruits are harvested before they reach full size. Therefore, pick them when their colour is bright. Once they loose their shine they are too old. Frequency of harvesting depends on the size of fruit. Small sized fruits are harvested more frequently than bigger or heavier fruits. The fruits after harvest are packed in baskets or cartons for local or distant market.

The average yield varies from 250 to 500q/ha of open pollinated varieties. While, in hybrids, it ranges from 300-700 q/ha under high fertility levels.

Storage of brinjal at room temperature for many days is not feasible. High respiration and water loss during storage affect the appearance and cooking qualities.Fruits packed in polythene bags and kept at room temperature can stay fresh for some more days. Keep them cool but not cooler than 9.9°C (50°F). Brinjal will spoil after about one week but will store nicely up to 2 weeks when refrigerated. Keeping quality of fruits varies with variety.

☆ The brinjal fruits are harvested when they are immature, the fruits are harvested when they reach marketable size.

☆ Although the fruit is harvested before it fully ripens, it should be allowed to attain a good size and colour.

☆ The fruits should be firm, and the outside colour glossy purple.

☆ Its surface should not lose its bright and glossy appearance.

☆ At harvesting, the calyx and stem-end are left attached to the fruit.

☆ Large, round varieties should be handled with care.

☆ Depending on the variety and the season it produces 250 to 400 quintals of fruits per hectare.

☆ Loss of glossy colour and dark coloured seeds are signs of over maturity.

☆ Brinjal fruits can be harvested until the first frost and should be picked as they mature to ensure continued fruit set.

☆ Harvesting is done by hand; the fruits are cut from the vines, with the calyx, or cap, and a short piece of stem left attached to each fruit.

☆ Fruits should be handled carefully because they bruise easily, which can result in significant surface disfiguration.

☆ Harvesting starts from 50th day onwards and continues for 50 days in the first phase and a second phase can be obtained after 20-25 days if adequate nutrition and irrigation are provided.

☆ Brinjal crop can be removed after 110 days if the 2nd flush is not desired, otherwise, it can be retained for 150–160 days.

☆ The size of the fruit reduces during the summer.

☆ The fruit can be stored for two to three days during winter and one to two days during summer under ordinary conditions but can be kept for about a week in a fairly good conditions at 7.2° to 10° C. and 85 to 95 per cent relative humidity.

☆ Brinjal fruits are sensitive to chilling injury below 50°F and deteriorate rapidly at warm temperatures, so they are not adapted to long storage.

After harvest of the crop, the remnants of the plant *viz*. straw, stables, leaves etc. are ploughed into soil to decompose, thereby providing source of organic matter for the next crop.

Purchasing Brinjal

Smaller, immature brinjals are best. Full size puffy ones may have hard seeds and can be better. Choose a firm, smooth-skinned brinjals that is heavy for its size; avoid those with soft or brown spots. Gently push with your thumb or forefinger, if the flesh gives slightly but then bounces back, it is ripe. If the indentation remains, it is over ripe and the insides will be mushy. If there is no give, the brinjal was picked too early. Also make sure that brinjal is not dry inside, knock on it with your knuckles. If you hear a hollow sound, do not buy it.

The quality criteria which decides the consumers acceptance are that the fruits should be free from blemishes, and have bright glossy appearance, firm in texture, light in weight and the calyx and stem should be fresh green. Brinjal fruits suffer from rapid loss of moisture leading to withering and loss of fresh appeal. The wrinkled skin on the other hand is a sign of excess loss from the commodity and adds to post harvest losses. The other factors which lead to losses are injury, diseases and pests. The estimated post–harvest losses in brinjal due to different diseases range from 3–7 per cent, which are mainly caused by *Alternaria tenuis, Helminthosporium speciferum* and *Trrichothecium* roseum. The most important pest deteriorating the quality is fruit borer *(Leucinodes orbonalis).*

Table 18: Standard for Grading the Vegetables for Marketing

Brinjal	General Grade	Commercial Grade
Long	No shrivelling, free from brown patches, green stalks	Shrivelling and brown spots, ≤ 20 per cent surface area, dull appearance
Round	No shriveling, proper shaped. No mechanical damage, lusturous, with green stalk, 65-150mm dia. & more than 5-10 per cent packing/ pressure damage, greenish brown area ≤ 10 per cent (Jan-March) & ≤ 25 per cent (Sept- Oct).	Up to 20 per cent shriveling, mis-shaped, dull, look dry or with out stalks, upto 20 per cent packing/pressure damage, greenish brown area ≤ 25 per cent (Jan- March) & ≤ 50 per cent (Sept- Oct).

Cooking

The raw fruit can have a somewhat bitter taste, but becomes tender when cooked and develops a rich complex flavour. Salting and then rinsing the sliced brinjal (known as degorging) can soften and remove much of the bitterness. Some modern varieties do not need this treatment, as they are less bitter. Brinjal is capable of absorbing large amounts of cooking fats and sauces, allowing for every rich dishes, but the salting process will reduce the amount of oil absorbed. The fresh fruit is smooth, as in the related tomato, the numerous seeds are soft and edible with the rest of the fruit. The thin skin is also edible, so that the brinjal need not be peeled. However, when young, the skin of most brinjals are edible but older ones should be peeled, since the flesh discolours rapidly and it should be cut just before using.

Storage and Relative Humidity

Brinjals are perishable and become bitter with age. They should be stored in cool, dry place and used within a day or two of purchase. To store in the refrigerator, place in a plastic bag. If you plan to cook it the same day you buy it, leave it out at room temperature.

Optimum Temp. Specific Heat* (C)	R.H. (%)	Storage Life (Weeks)	Temp. Limit –	Chilling Injury Symptoms	Water Content (cal/g/ºK)
10.0–11.0	95%	3–4	7	Surface sealed	93

*: See Ryall and Lipton (1979).

Chapter 10
Brinjal Seed Production

The increased activity in vegetable production have led to an interest in the producton of vegetable seeds. Vegetable seed is an important commodity because the degree of quality can ultimately have a major impact on yields. However, producing high quality seeds is a challenge in itself- it involves carefully controlled factors as well as combining the most desirable traits from various different varieties. Seed production requires much more time and technology than simply growing plants for fresh markets. This is seen as an area of great potential that could prove to be boon to the farmer and a significant earner for the country. All parts of the country provide the ideal climate and condition for profitable vegetabe seed production. Seeds are produced to ensure that farmers can grow high quality vegetable crops. At present a limited number of farmers are producing vegetable seeds. One major reason for this is that they simply lack the technical know how. The Indian seed industry at present is worth about 500 crores of which around 200 crores is export market. Private sector is contributing 60 per cent by value and 40 per cent by volume in organized seed sector. If we see the opportunities, the Indian seed market can cross more than 1000 crores worth of business in coming 10 years (Abid Hussain, 2003).

Land Requirement and its Preparation

The land for brinjal seed production should be free from volunteer plants. The soil of the selected field preferably be sandy loam, fertile, friable, rich in organic matter and well drained. Land should be prepared to a good tilth by ploughing and harrowing 4–5 times before transplanting the seedlings. Bulky organic manures should be incorporated into the soil during land preparation. When land is sufficiently prepared and leveled, the whole plot is divided into the beds of suitable dimensions before transplanting.

Time of Sowing

In hills it is March–April, where as in Plains it starts from February to March, June to July and October to November.

Seed (500g/ha)

The seed should be nucleus/breeder or foundation, procured from a reputed and certified agency.

Nursery Raising

Raised nursery bed is prepared for raising brinjal seedlings. Top soil of the nursery bed is mixed properly with well rotten farm yard maure or compost or leaf mould. The seeds treated with 30 ppm IAA are sown at a depth of 5 to

10 mm depth in rows of 5 cm apart. The seeds after they are sown are covered with a fine layer of soil and compost. The nursery bed is then covered with locally available mulching material like paddy straw or dry grass like *Sachharum spontanium* till seeds germinate. Usually seeds germinates within 12 to 18 days of sowing.

Transplanting

Under good nursery management, brinjal seedlings become ready for transplanting in about 20 to 25 days after sowing. The seedlings should be hardened by withholding watering for 4 to 6 days before transplanting. However, nursery bed should be softened with water, if soil becomes too hard, so that roots are not injured when the seedlings are uprooted. Crumbling of roots should be avoided while transplanting.

The distance of transplanting for non-spreading brinjal varieties would be 50 to 60 cm both ways and for spreading varieties, the row to row distance should be 75 to 90 cm and for plant 60 to to 70 cm (Singh, 1989).

Transplant the seedlings when 12–15 cm tall, preferably at evening time. Irrigate immediately afterwards.

Manure and Fertilizer

Brinjal seed crop is comparatively long duration crop. For better fruiting and seed yield, application of balanced nutrients is essential. They are supplied both through organic manures and fertilizers.

Apply 250 to 300 quintals of farm yard manure per hectare at the time of land preparation and 60 kg/ha nitrogen, 80kg/ha phosphorus and 60kg/ha potash at the time of transplanting. Top dress 60kg/ha nitrogen at two splits. The first top dressing should be done 15 days after transplanting and the second at flowering time. Irrigate the crop immediately after top dressing.

Method and Time of Application

Both method and time of application of nutrients is important for successful brinjal seed production. Normally one fourth of the total nitrogen alongwith the whole of phosphorus and potash should be applied within 24 hours before transplanting of seedlings. The half dose of remaining quantity of nitrogen should be applied 25 to 30 days after planting and second application during flowering (45 to 50 days after transplanting). Top dressing of nitrogen fertilizer is done in two ways (i) band application *i.e.* fertilizer is placed either in shallow furrow opened along the row of brinjal plants or (ii) fertilizer is placed on ground along the row of the plants and shallow digging suffice mixing of it. Top dressing can also be done by foliar spraying. The concentration of solution should not exceed 2 per cent otherwise nitrogen may cause leaf scorching or leaf burning. Foliar application is also advantageous, as one can also apply pesticides and fungicides simultaneous thereby reducing cost of cultivation.

Irrigation

Immediately after transplanting, light irrigation is very essential for survival and establishment of the seedlings. Subsequent irrigations are needed according to season. During summer, brinjal crop requires irrigation at very fourth or fifth day whereas at 10–12 days interval during the winter. Higher yields are obtained when crop is irrigated at right time of its requirements. Timely Irrigation is essential especially for fruit set and development. Drip irrigation if available saves 30-40 per cent water.

According to Balkrishnan *et al.,* 1993, brinjal crop can be managed to prevent from moisture stress by seed soaking in 1.0 per cent KH_2PO_4 solution for 12 hours. The crop is given stress for about 15 days during flowering by withholding irrigation, and such stressed crop sprayed with 1.0 per cent KCP, significantly recorded higher yields because sprayed plants maintained higher photosynthesis (CU2/hrx, proline content (160U2 Mg/g) coupled with low transpiration rate (1.97 Mg H2/cm²/sc) which might have led to higher yields.

Interculture and Weed Control

It is better to keep the field free of weeds especially at initial stages of crop growth and is done by two to three light hoeings and earthing up. Therefore, during early stage of crop growth shallow cultivations are done to provide good environment for better root development and will also help in keeping down the weeds This facilitates better aeration to root system and gives support to plants. Besides, aggressive weeds are controlled with use of herbicides. Application of fluchloralin (Basalin) @ 1.5kg a.i./ha as a pre-emergent weedicide, applied one week after transplanting seedlings, followed by one hand weeding at 30 days after planting controls a broad spectrum of weeds. Use of black polythene mulches is also efficient for suppression of weeds and for better growth of plants. Weeds commonly seen growing with brinjal crop are like Amaranthus viridis, *Acanthospermum hespidurm, Cyperus rotundus, Digitaria marginata, Orobancha* spp. *Portulaca oleracea, Trianthema portuascastrum* etc.

Mulching

Use of black plastic or black polythene mulch in brinjal helps to stimulate growth and improves earliness. Mulch ultimately reduces the infection rates of verticilium wilt and increases the brinjal yield. The other mulch materials which are locally available may also be used *viz.* Guinea grass, *Panicum maximum* at 4 tonnes per hectare.

Chapter 11

Isolation Requirements

Brinjal is partially self and cross pollinated, but self pollination is more common and varying amount of cross pollination takes place because of heterostyly. Cross pollination is mainly through honey bees and bumble bees. The extent of natural crossing depends upon insect activity and has been recorded from 0–48 per cent. For seed to be pure, the an isolation distance of at least 200 meters for foundation and 100 meters for certified is recommended. The field inspection should be done at (a) before flowering (b) flowering and fruiting stage and (c) mature fruit stage and prior to harvesting. Leaving one or two harvests for vegetable purpose is advisable for detection and removal of off typs and to avoid chances of contamination from off types.

Contaminants	Minimum Distance (meters)	
	Foundation	Certified
Fields of other varieties	200	100
Fields of the same varieties	200	100

Specific Requirement

Factors	Maximum	Permited (%)#
Off types	0.10	0.20
Plants*affected by seed borne diseases	0.10	0.50

\# Standards for off types shall be met at and after flowering and for seed borne diseases at final inspection.

* Seed borne diseases shall be: phomopsis blight (Phomopsis vexans).

Roguing

Roguing of off type plants much before blossoming is essential. While the first fruit is still only partially developed, it is possible for growers familiar with varieties to rogue out more off type plants. Later, when each plant has several more or less mature fruits, rouging can be based on fruit size, shape and colour as well as the plant over all performance. In addition to off types, plants affected by disease such as phomopsis blight and little leaf etc. should be removed time to time from the field. Therefore, to maintain genetic purity of a variety, it is imperative on the incumbent to carryout the rouging operation carefully. At least minimum of three roguings will suffice the purpose. The roguings are done at three different stages of crop growth.

1. Before Flowering

In this rouging, the plants which are in flowering process or have already borne flowers are removed. The plant showing different type of branch orientation, leaf colour, presence or absence of pubescents should be removed, if they are different than grown variety. Plants may be erect type and or spreading type but when such plants show differences from the variety under cultivatiuon, are removed. Also plants affected by known or unknown diseases or disorder should be removed.

2. At Flowering and Fruiting

The second rouging is done at the time of flowering and fruiting stage. In this rouging, the plants which have not flowered or the plant which are in flowering process should at once be removed. The plants showing different flower colour, flower orientation and leaf colour should be removed if they are different than the grown variety. At the time of fruiting stage, plants showing different fruit colour, fruit shape and those plants which are flowered should be removed if they are different than the grown variety. Also plants affected by disease and insect-pests should also be removed from the field.

3. At Mature Fruit Stage and Prior to Harvesting

The third and the final rouging is done at the time of maturity. In this rouging, the plants which are having immatured or overmatured fruits should be removed from the field. At this stage fruits for colour, shape, size and other external features such as phomopsis blight, little leaf or bacterial wilt etc. are thoroughly checked and affected plants if any should be removed from the field and burnt.

Harvesting and Thrashing

The fruits are allowed to become mature, to ripen in the plant itself and the ripe fruits appearing shining yellow in colour are picked and collected. In the ripe fruits an absecission layer develops between the proximal end of the fruit and the calyx resulting in fruit fall on the ground; such fruits are collected for seed extraction. It is not necessary to wait the fruits to fall on the ground and even before they can also be harvested. After the fruits are harvested, their outer covering is peeled off and the flesh along with the seed is sliced. Fruits usually ripe in 75–80 days in early varieties and 110 to 125 days or so in late varieties after transplanting. The crushed or sliced fruits are soaked overnight in baskets for softening. This results in easy separation of seeds from the pulp when pulp is stirred next day morning. If the material is allowed in this condition to stand overnight, the separation of seeds from the pulp becomes easier.The brinjal seeds are minscule and not easily separated from the flesh. If you have waited long enough, the pulp can sometimes be teased away from the seeds by hand. If that does not work, pulverize it with a potato masher or with a small food processor, so that the maximum number of seeds are exposed. In other method, seed extraction is achieved when commercial HCI is added in pulp at 1:4 ratio (25 ml HCI/kg pulp). The pulp is stirred for 25 to 30 minutes and then seeds are separated out by washing and sieving. After separation, the seeds are washed with running water. This is accomplished by put the entire mess in a pail of tepid water and leave it there until the seeds have settled to the bottom. Those, which float, such seeds are either dead or have very poor germinability, should be rejected. Skim the top off and pour the remains through cheese cloth or a fine stainer. Some times it needs several rinses to acive a state of marginal cleanliness, but do not feel that the final product must look as sparkling as that the seed man provide. No fermentation process is given. If the seeds are not dried quickly *i.e.* 12 hours, they may sprout. After washing, the seed is spread thinly to dry in partial shade to a moisture content of 8 to 10 per cent or below, before storing. The recovery of seeds is generally higher from medium sized fruits. Quality seeds can be recovered in about 8 pickings.

Some times fruits are pulped in an electric pulper and treated with commercial HCl at 1:40 ratio (25cc per kg of pulped fruit), stirred for 25 to 30 minutes and washed well and dried to 8 to 10 per cent moisture content. Seeds are attractive and free from fungal attack. A maximum seed recovery of 5 per cent to the total weight of fruits can be registerd by acid method. The dried seed is than treated with 4g Captan or Thiram per kg before storing it. It can be stored in paper bags for about one year and up to three years in aluminium foils.

Seed Yield

It depends upon variety used, climatic conditions, plant density and other agronomical packages followed. The highest yield is obtained at the close spacing of 60 × 60 cm (Singh and Syamal, 1995). The 1000 seed weight is approximately 4 to 5 g. On an average seed yield varies from 360–400 kg/ha, while as Hybrid seed yield is 50–60 kg/ha.

In Arka Neelkanth variety, the smaller sized fruits (15 g) were found to contain minimum quantity of seeds per 100 g fruit (2.46g) while larger sized fruits (120 g) had it maximum (4.752g/100g fruit). The test weight of the seeds (1000 seed weight) was maximum (6.130g) with smaller fruits. The germination percentage of seeds obtained from the smaller fruits was substantially lower (67 per cent) as against 95 per cent with seeds from the larger fruits. In another study it has been found that seed content and seed quality vary with different four portions of brinjal fruit (cv. Arka Neelkant). The second half of the fruit as against 27 per cent in the first half of the fruit (on the stalk side), contained over 73 per cent of the total seeds of the fruit as against 27 per cent in the first half of the fruit (on the stalk side). The seed content is the least in the first quarter (50 per cent). The second and the third quarters contain the maximum quantity of seeds (26 and 52 per cent respectively). The last quarter contains 17 per cent of the total seeds. The test weight of the seeds has been maximum in the seeds from the first quarter (5.48g/100 seeds) and it decreased successively till the third quarter and at par with fourth quarter. There has not been much variation in the germination and vigour of seeds among the first three quarters. While the seeds from the last quarter having less germination as well as vigour (Naik *et al.*, 1993b).

Packing and Storage

Dried seeds are packed in glass jars, polythene bags or plolythene boxes. The seeds are Also packed in cloth bags or in tinwares. Packed seeds are stored in dry clean and air tight containers in cold storage at 10°C.

Production of Hybrid Brinjal

F1 hybrids are high yielders, better quality in respect of uniformity in fruit size, shape, taste, maturity and high degree of tolerance against pests and diseases. The cost of F1 seeds are costly, but the production technique is easy to cross between two parental lines through hand emasculation and pollination. A fruit so obtained after pollination contains large number of seeds. Moreover, the seed requirement per hectare is only 400- 500 g. Thus the cost of hybrid seed production is not so high.

Both the lines (male and female) are seen carefully and off types if any are strictly removed. Emasculation is done one day in advance to anthesis. It occurs between 7 a.m to 11 a.m. Maximum pollen grains are available at 9.30 to 10.30 a.m and stigma is most receptive at the time of opening of flower. The pollen is collected and placed properly on the receiptive stigma with the help of a brush. Pollinated flowers are bagged and unpollinated flowers are removed. The best period is October and November. The success of fruit setting is more in fore-noon pollination as compared to afternoon pollination.There is positive correlation between fruit setting and pollination in fore-noon and negative correlation between fruit setting and wind velocity (Mandal *et al.,* 1995). Fully ripe fruits are harvested and all operations done accordingly as mentioned in the *fore going text.* Following precautions are needed to be taken *viz:*

Emasculated flowers should be pollinated soon.

Bagging and tagging are helpful in confirming the fruits obtained from hybridization.

Always avoid cloudy and windy weather during pollination.

Table 19: Field Standards for Brinjal Seed Production
(maximum permitted in per cent)

Factors	Foundation	Certified	Remarks
Off types	0.10	0.20	*Strict roguing for phomopsis blight and little leaf diseases
Other types	—	—	—
Objectionable weed plants	—	—	—
Disease plants*	0.10	0.50	—

Seed Standards

Factors	Standards for Each Class	
	Foundation	Certified
Pure seed (minimum)	98.0 per cent	98.0 per cent
Inert matter (maximum)	2.0 per cent	2.0 per cent
Other crop seeds (maximum)	None	None
Wee seeds (maximum)	None	None
Germination (minimum)	70.0 per cent	70.0 per cent
Moisture (maximum)	8.0 per cent	8.0 per cent
For vapour proof containers (maximum)	6.0 per cent	6.0 per cent

Class	Genetic purity (%)	(minimum)*
Certified	90.0	

*During grow out test, off type plants (other selfed plants) such as segregants, outcross and plants of other hybrids should not exceed more than 1.50 per cent out of the 10 per cent plants earmarked for selfed plants.

The minimum population size of 400 plants shall be maintained in row, replicates of 200 each or four of 100 each throughtout the test and each plant shall be examined individually.

Factor	Genetic Purity (%)	Minimum Reject Number
Certified	90.0 (10 in 100)	44

Seed Production of Brinjal

☆ The practices normally followed for the crop raised for fruit production need to be adopted for seed production.

☆ It is a self pollinated crop but cross pollination may also take place.

☆ The two varieties meant for foundation seed production may be kept 200m apart to avoid contamination.

☆ For the purpose of seed production the ripe fruits turn yellow are crushed and stored over night. The seed after washing with water is sieved and dried.

☆ The washing is usually done in the morning so that the seed is at least half dried during the day otherwise it may germinate.

☆ Seed yield of Brinjal is 590–880 kg/ha.

Requirement and Availability of Brinjal Seed

Vegetable	Requirement (Tonnes)	Seed Availability (Tonnes)	Replacement (Per cent)
Brinjal	239.54	151.87	63.40

Source: Seed Association of India.

Disease and Pests

Same as described for general crop.

Bibliography

Abdelfallah, M.A and A.S. Abdel Salam. (1972). Hort. Abstr. 45 (11): 8507.

Aheer, G.M., Saeed,M., Latif, M. (1996). Efficacy of insect growth regulators against brinjal fruit and shoot borer (*Leucinodes orbonalis*). Second International Congress of Entomology Sciences, Islamabad (Pakistan). 19- 21, March,1996.Parc,1996.p75.

Anonymous. (2000) please see Advances in vegetable production.

Aliev, D.A. (1968). Trudy azerb nauci ssled inst. o VOSC.2: 157-68

Aliev, B. (1973). Hort. Abstr. 43 (6): 377.

Baily, L.H. (1949). Manual of cultivated plants. 2nd Ed. Macmillan Company, New York.

Baha-Eldin, S.A and Blackhurst, H.T (1968). Proc. Amer. Soc.Hort. Sci. 92: 480-Bhaduri, P.N. (951). Indian J. Genet. 2: 75-86.

Bhaduri, P.N. (1932). Journal of Indian Botanical Society. Vol. 11. p.202-224.

Burkill, I.H. (1935). A Dictionary of the Economic Products of the Malay Peninsula, Crown Agents for the colonies, London, p. 2402.

Chadha, M.L. 1993. Improvement of brinjal. In K.L. Chadha and G.Kalloo (eds.). Advan. in Horticulture Vol. 5-6 Vegetable Crops Part 1, p.105-135. Malhotra Publ. House, New Delhi.

Chakraborty, A.K and Choudhury, B. (1975). Proc. Indian Nat. Sci. Acad.B, 41: 379-85.

Chandrasekaran, P and C.M. George. (1973). Agri. Res. J. Kerala. 11 (2): 106-108.

Chen, N; Li, H; Kalb,T. (2001). Suggested cultural practices for egg plant. AVRDC Training Guide. Asian Vegetable Research Development Center. Shanhua, Taiwan.

Choudhary, B. (1976). Vegetables (4th edn.) National Book Trust, New Delhi p.50-58.

Choudhary, B. (1976). Evolution of crop Plants. Ed. N.W. Simmonds, Longman Inc, London and New York,p.278-9.

Choudhary, B. (1966) Indian Hort.10: 56-58.

Decandole, A. (1886). Origins of cultivated plants.

De Candole, E. (1959). Origins of cultivated plants. Hanger Publishing Company, New-York.

Decker (1951). Phytopathology, 41: 9

Deshpande, A.A., Bankapus, V.M and Nalawadi, U.G. 1978. (1978). Current Research, University of Agricultural Sciences, Bangalore, vol. 7. p.174.

Dhankhar, B.S., Gupta, V.P and Sing, K. (1977). Haryana J. Hort. Sci. 6: 50-58

Doijode, S.D.2001. Seed Storage of Horticultural Crops. Haworth Press. ISBN 1560229012

Ellis, B and Bradley, F. (1996). The organic gardener's hand book of natural insect and disease control. Rodale press. Emmaus, Pennsylvania.

Gibbon,D. (1973). J. Hort. Sci. 48 (3): 217-21.

Gavaskar, D and Anburani, A. (2004). Influence of plant growth regulators on growth attributes in brinjal (*Solanum melomgena* L.) cv. Annamalai. South Indian Horticulture.Vol.52: (No.1-6).

Gill,H.S., Arora, K.S and Pachauri, D.C. (1978) Indian J. Agric. 46: 484-90.

Gnanakumari and G. Satyanarayana. (1971). Indian J. Agri. Sci. 41 (6): 554-58.

Gopalan, G. Rama Sastri, B.V. and S.C. Balasubramanian (1995). Nutritive Value of Indian Foods. National Institute of Nutrition. I.C.M.R, Hyderabad, India.

Gowda, D.M.V. (1977). Mysore J. Agric. Sci. 11 : 426.

Gupta, S.S., Kalda, T.S., Singh, Narendra, and Kumar, Ravinder. (2000). Indian Hort. Vol. 45. p.17-19.

Gupta, S.S., Kalda, T.S, Narendra, Singh and Kumar, Ravindra. (1999). Indian Hort. Vol. 44, p.7-8.

Hendricks, T and L.C.Van Loon (1990). Petunia peroxidase is localized in the epidermis of aerial plant organs. J. Plant Physiol. 136: 519-25

Jayaram, K.M and Neellakandan, N. (2000). Effect of plant growth regulators on sex determination in *Solanum melongena* L. Indian J. Plant Physiology. Vol: 5, Issu: 3

Jotishe, R and Chandra, P. (1969). Hort. Abstr. 41 (2): 4160

Kakizaki, V. 1924. The flowering habit and natural crossing in egg plant. Jpn. J. Genet. 3.29.

Kakizaki, Y. 1930. Breeding crossed eggplants in Japan. J. Hered.21.253.

Kalda, T.S and Gupta, S.S. (1990). Indian Hort. Vol.41. p.4-5.

Kalda, T.S., Swarup, V and Choudhury, B. (1976). Veg. Sci., 3: 65-70.

Kalda, T.S., Swarup, V and Choudhury, B. (1977). Resistance to Phomopsis blight in egg plant. Vegetable Sci. 4 (2): 90-101. (Scientific Hort. Vol.1-1991.

Kalloo, G.1993. Egg plant- *Solanum melongena* L. In G. Kalloo and B.O. Bergh (eds.). Genetic Improvement of Vegetable Crops.

Kalloo, G.1988.Vegetable Breeding. Panima Educational Book Agency. Vols. 1-11-111,p.64.

Kalloo, G., Baswana, K.S and Sharma, N.K. (1993). Indian Hort. Vol.38. p.12-13.

Kalloo, G., Dudi, B.S., Singh, Amir and Salyan, D.S. 1993. Indian Hort. Vol.38. p.11-12

Kushal, N and Sudha, S.K (1995). Role of Phomopsis vexans in damping off of seedlings in egg plant and its control. Indian J. Mycol. & Plant Pathol. 25 (3) 189-191.

Lal, S., Verma, G and Pathak, M.M. (1974). Indian J. Hort. 36: 51-55

Lenz, F. (1970). Hort. Res. 10: 81-82.

Lindgren,N. (1982). http://laurpkbs. Unl. edu/ horticulture/g 603.htm.

Lombardi, D. and Restaino, F. (1981). Colture Protette. 10: 31-36

Mandal, S.C., Singh, Y.V. and Harihar Ram (1995). Hybrid seed production in brinjal (*Solanum melongena* L.) under Tarai coditiuons of U.P. Nat. Sym. On Recent Dev. In Vegetable Impr. Held at Indra Gandhi Krishi Vishwa Vidyalaya, Raipur (M.P) on February 2-5.

Martin, F.W. and Rhodes, A.M. (1979). Euphytica,28: 367-83.

Meena, S.S., Dhaka, R.S and Jalwania,R. (2006). Economics of PGR's in brinjal (Sol-L) under semi-arid conditions of Rajasthan. Indian Society of Agri. Sci.Vol.25.Issue. 4.

Meena, S.S and R.S. Dhaka. (2003). Effect of plant growth regulators on growth and yield of brinjal under semi-arid conditions of Rajasthan. Annals of Agricultural Resh. Vol. 24 (No.3).

Mishra, G.M. (1961). Indian J. Hort. 19: 305-17.

Naik, L.B., Prabhakar, M and Doijode, S.D. (1993 b). A study on quality and content of seeds in different portions of fruit in brinjal. Golden Jubilee Sym. Hort. Res. Changing Scennrio held on May 24-28 atb iihr, Bangalore (India.

Nandkarni, K.M. (1927). Indian Materica Medica, Nadkarni and Sons, Bombay.

Nothmann, J. and Koller, D. (1973). Israel J. Botany.22: 231-5

Nothmann, J., Ryiski, I. and Spigelman, M. (1979). Scientia Hort. 11: 217-20.

Pal, B.P and Singh, H.B.1943. Floral characters and fruit formation in the egg plant. Indian J. Genet. Plant Breed. 3.45.

Pal, B.P and Singh, H.B. (1946). Indian J. Genet. 6: 19-33

Pal, B.P and Singh, H.B. (1949). Indian Fmg. 10: 378-80.

Parthasarathy, (1948)

Patil, Shivashankargouda B., Merwade, M.N., Vyakaranahal, B.S. (2008). Effect of growth regulators and fruit load on seed yield and quality in brinjal hybrid seed product. Indan J. Agri. Resh. Vol:42: issu 1

Peter, K.V. and Singh, r.d. (1973). Indian J. Agric. Sci. 43: 452-5.

Popova, D.1958. Some observations on flowering, pollination and fertilization of the egg-Plant. News. Inst. Plant. Industry, Sofiya. 5.211

Prasad, D.N and Prakash, R.1968. floral biology of brinjal (Solanum melongena L.). Indian J. Agri. Sci. 38.1053.

Purewal, S.S. (1957). Vegetable Cultivation in North India, Farm Bull. I.C.A.R. New Delhi, No. 36.

Rangaswamy,P and Kadambavanasundaram, M (1973). South Indian Hort. 21: 21-6.

Rao, M.V.B., Sohi, H.S and Vijay, O.P. (1976). Veg. Sci. 3 : 61-64

Roxburgh, W. (1932). Flora India. Vol.1, p.153-163

Ruiter, D.E. (1974). Hort. Abstr. 45: (6): 4107.

Sambandam, C.N. (1964). Indian J. Genet. 24: 175-6.

Sambandam, C.N. (1969). Annmalai Univ. Agric Res. Ann.1: 1-7.

Sampson, H.C. (1936). Bull. Misc. Inf. Roy. Bot. Gdn. Kew, Add. Ser.12:1.7.

Sampson, H.C. (1936). Bull. Misc. Inf. Roy. Bot. Gdn. Kew. Add. Ser. 12:159

Seth, J.N and D.G. Dhauder. (1970). Prog. Hort. 1 (4): 45-50.

Sharma,M.D. (2006). Effect of plant growth regulators on growth and yield of brinjal at Khajura, Banke. J. Inst. Agri. Animal. Sci.27: 153-156.

Shukla, V. and Naik, L.B. (1993). Agro-techniques of Solanaceous Vegetables. Advances in Hort. Vol.5, Vegetable Crops, Part 1 (K.L.Chadha and G.Kalloo, eds.) Malhotra Pub. House, New Delhi. P. 365.

Sidhu, A.S., Kour, G. and Bajaj, K.L. (1982). Veg. Sci. 9: 112-8

Singh, B., Joshi,S. and Kumar, N. (1978). Haryana J. Hort. Sci. 7: 95-99.

Singh, K and D.S.Sandhu (1970). Effect of soil and foliar application of nitrogen on growth and yield of brinjal (*Solanum melongena* L.) Punjab Horticulture. 10: 103.

Singh, S.P. (1989). Production technology of vegetable crops. Agriculture Research Communication Centre, Karnal, Haryana, India.

Singh, V.N and Syamal, M.M (1995). Effect of nitrogen and spacing on yield and quality attributes of brinjal (*Solanum melongena* L). J. Rs. Birssa Agri. Univ.7 (2): 137-139.

Som, M.G and Mallik, S.C. (1979). Orissa J. Hort.7: 28-32.

Srinivasan, K and Gopimony, R. (1969). Agri. Res. J. Kerals. 7: 39-40

Sridhar, S; Arumugasamy,S; Saraswathy,H; Vijaya Lakshmi, K. (2002). Organic Vegetable Gardening Center for Indian Knowledge systems, Chennai.

Srivastava, B.P., Srivastava, J.P and K.P. (2000). Indian Horticulture. Vol.44, p.36

Swaminathan, M and Srinivasan,, K. (1971). Agri. Res. J. Kerala. 9: 11-13.

Swamy Rao,T. (1970). Madras Agric. J. 57: 508-9.

Tanckev, S.S., P.J. Ruskova and C.F. Tcaiberlake. (1970). Phyto chem.. 9: 1681-82.

Thakur, M.R., Singh, K. and Singh, J. (1968). J.Res. Punjab Agric. Univ. Ludhiana. 5: 490-95.

Thakur, M.R., Singh, K. and Singh, T. (1969). J. Res. Punjab Agric. Univ.Ludhiana. 6: 769-75.

Thompson, C.H. and Kelly, C.W. (1957). Vegetable Crops. Mc Graw-Hill BOOK Co. Inc. New York.p.501

Tsao and Lo 2006 In Y.Hui, Hand Book of Food Science, Technology and Engineering. B O CA Rato: Taylor and Francis.

Umrani, N.K and B.D. Khot. (1973). Indian Agri Sci. 43 (8): 876-8.

Umrani and Khot (1974).

Vavilo, N.I. (1928). Proceeding of 5th. International Congress of Genetics, New York, p. 342-69.

Verma, T.S., Gill, H.S., Joshi,S and Atma Ram. (1992). Indian Horticulture, Vol.3,p.8-9.

Viswanathan, T.V. (1973). Proc. Indian Acad. Sci. 77. (4): 176-80

Viswanathan, T.V. (1975). Current Science. 44:134.

Yadav, B.S., Nandwana, R.P., Lal, A. and Verma, M.K. (1975). Indian J. Mycol. Pl. Path. 5:

Zeven, A.C. and Zhukovsky, P.M. (1975). Dictionary of cultivated Plants and their Centers of Diversity, Wageningen, Netherlands, p. 219.

Index